Springer Theses

Recognizing Outstanding Ph.D. Research

For further volumes:
http://www.springer.com/series/8790

Aims and Scope

The series "Springer Theses" brings together a selection of the very best Ph.D. theses from around the world and across the physical sciences. Nominated and endorsed by two recognized specialists, each published volume has been selected for its scientific excellence and the high impact of its contents for the pertinent field of research. For greater accessibility to non-specialists, the published versions include an extended introduction, as well as a foreword by the student's supervisor explaining the special relevance of the work for the field. As a whole, the series will provide a valuable resource both for newcomers to the research fields described, and for other scientists seeking detailed background information on special questions. Finally, it provides an accredited documentation of the valuable contributions made by today's younger generation of scientists.

Theses are accepted into the series by invited nomination only and must fulfill all of the following criteria

- They must be written in good English.
- The topic should fall within the confines of Chemistry, Physics, Earth Sciences, Engineering and related interdisciplinary fields such as Materials, Nanoscience, Chemical Engineering, Complex Systems and Biophysics.
- The work reported in the thesis must represent a significant scientific advance.
- If the thesis includes previously published material, permission to reproduce this must be gained from the respective copyright holder.
- They must have been examined and passed during the 12 months prior to nomination.
- Each thesis should include a foreword by the supervisor outlining the significance of its content.
- The theses should have a clearly defined structure including an introduction accessible to scientists not expert in that particular field.

Sara Bobone

Peptide and Protein Interaction with Membrane Systems

Applications to Antimicrobial Therapy and Protein Drug Delivery

Doctoral Thesis accepted by
the University of Rome Tor Vergata, Italy

Springer

Author
Dr. Sara Bobone
Department of Chemical Sciences and Technologies
University of Rome Tor Vergata
Rome
Italy

Supervisor
Prof. Lorenzo Stella
Department of Chemical Sciences and Technologies
University of Rome Tor Vergata
Rome
Italy

ISSN 2190-5053 ISSN 2190-5061 (electronic)
ISBN 978-3-319-06433-8 ISBN 978-3-319-06434-5 (eBook)
DOI 10.1007/978-3-319-06434-5
Springer Cham Heidelberg New York Dordrecht London

Library of Congress Control Number: 2014937923

Printed on acid-free paper

Springer is part of Springer Science+Business Media (www.springer.com)

Supervisor's Foreword

Membranes are fundamental elements of the cell, and their properties are dictated by the amphiphilic character of the lipid molecules composing them. Membranes form by *self-assembly*: in water, phospholipids spontaneously arrange into a bilayer, in order to shield their aliphatic tails from the polar solvent. The hydrophobic effect holds lipids together, but at the same time allows a high degree of mobility, so that a lipid bilayer is a highly dynamic environment. Membrane *fluidity* is essential for vital cellular functions, such as mobility, exo- and endocytosis, fusion, etc. Finally, due to their hydrophobic core, lipid bilayers exhibit *selective permeability* to solutes, thus regulating the contents of the cell. Any breach in the barrier properties of the membrane inevitably leads to cell death.

Thanks to the self-assembly properties of lipids (and phospholipids in particular), artificial membranes are easily formed. One of the most popular model system used to mimic biological membranes are unilamellar liposomes, i.e. vesicles constituted by a single lipid bilayer. In these very controlled and reproducible systems lipid composition can be varied at will.

In her thesis, Dr. Sara Bobone exploited liposomes both as mimics of cellular membranes, to understand the mechanism of action of a class of bioactive molecules called antimicrobial peptides, and as nanocontainers for the synthesis of nanoparticles for enzyme-replacement therapy.

Antimicrobial peptides (AMPs) are natural oligopeptides endowed with strong antibacterial activity, exerted principally by perturbing the permeability of microbial membranes, thus leading to the dissipation of transmembrane gradients and to the loss of cellular contents. Thanks to their relatively aspecific target, and to their fast bactericidal action, AMPs represent a promising source of lead compounds to address the global problem of bacterial resistance to presently available antibiotic drugs. Unfortunately, their mechanism of membrane perturbation, and the peptide properties responsible for their activity and selectivity, are still a matter of debate. This lack of detailed knowledge presently limits the development and clinical applicability of AMPs-derived molecules.

The same fluidity, which is essential for the vital membrane functions, makes the structural characterization of peptide-membrane systems by the most powerful approaches (X-ray crystallography and NMR) particularly difficult. In this thesis, by employing alternative techniques (fluorescence spectroscopy and neutron reflectivity, coupled to MD simulations), Dr. Sara Bobone tackled several conundrums regarding AMPs that had been debated in the literature for years.

First, she showed that the fluidity of membranes allows them to adapt to peptide insertion in the bilayer, so that even a peptide much shorter than the normal membrane thickness can be able to reach from one side of the bilayer to the other, thus forming pores.

Then, she addressed the issue of AMPs selectivity, i.e., the ability of these molecules to discriminate between prokaryotic and eukaryotic membranes, thanks to their different composition. Peptide charge was long considered as the most important property under this respect, due to the much higher concentration of anionic lipids present in the outer layer of the bacterial plasma membrane. However, the work presented in this thesis shows that this is not enough, and that conformational effects on peptide hydrophobicity play a very important role, too.

Finally, she showed that subtle modifications to peptide sequence and properties can profoundly change activity (e.g., from antimicrobial to cell-penetrating). Her findings indicate that different bioactive peptide classes are just different manifestations of the same membrane-active behavior.

Overall, these findings provide an important starting point for the rational design of a new generation of AMPs-inspired antibiotic molecules.

In the second part of the thesis, liposomes were used for a completely different purpose. Several diseases are caused by the loss of function of a particular enzyme, and therefore exogenous supplement of that protein would be an obvious therapeutic approach. However, direct application of proteins as drugs is complicated by their susceptibility to degradation and immunogenicity. Dr. Bobone synthesized novel nanodevices, in which an enzyme is entrapped inside hydrogel nanoparticles, whose mesh size prevents the protein from being released, but allows diffusion of substrate and products. Liposomes were used as nanotemplates for the synthesis of monodisperse polyacrylamide hydrogel nanoparticles under nondenaturing conditions. This system protects the protein from antibodies and degradation, while allowing it to exert its catalytic function. Biocompatible, size-tunable nanoparticles have been obtained, with a good entrapping efficiency, and conservation of enzyme activity.

In my opinion, this thesis presents several novel and relevant findings, with important applications to the rational design of new antibacterial drugs, and for the realization of nanoparticles for the delivery of proteins in therapeutic applications. Apparently, this view was shared by the Italian Chemical Society, which awarded to Dr. Bobone the Semerano Prize for the best physical chemistry thesis in the years 2012–2013.

The introductory sections and the extensive reference list also represent a good starting point for novices to the fields of artificial membranes, antimicrobial peptides, and protein drug delivery.

Rome, March 2014 Prof. Lorenzo Stella

Preface

The plasma membrane is a fundamental element of the cell, constituted mainly by phospholipids, sterols, and membrane proteins.

Biological membranes carry out the function of a barrier between the living cell and the environment and between intracellular compartments and the cytosol. The lipid bilayer regulates the transport of molecules and their concentration inside cells. It is a semipermeable structure, which allows the entry of small hydrophobic molecules, like oxygen and carbon dioxide. Small polar molecules, like water and ethanol, can cross the bilayer, but they do it more slowly. By contrast, the diffusion of charged ions, water-soluble molecules, or large molecules is avoided. The passage of these substances is regulated by specific transport proteins, selective channels for ions and molecules, or pumps capable to actively transfer molecules across the bilayer. Other transmembrane proteins are receptors for extracellular signaling molecules, which are critical for controlling cellular properties and behavior, like hormones, neurotransmitters, growth factors. Other receptors are enzymes, whose role is to start a metabolic cascade of intracellular reactions when stimulated. Furthermore, some proteins, or protein groups, are disposed in the membrane in a definite order, to create a system of electron transfer and energy accumulation in the form of ATP.

When the structure of the lipid membrane is damaged, the intracellular content is released outside the cell, causing rapidly its death. This can be exploited to kill bacterial cells using membrane-damaging agents. A possibility is represented by antimicrobial peptides (AMPs), which are able to perturb bacterial membranes, altering their selective permeability and leading bacteria to death.

In this thesis, the study of peptide-membrane interactions has been carried out using model membranes. It has been shown, indeed, that peptides feature the same activity on these systems. The most common membrane model is represented by lipid vesicles (liposomes), a simple but realistic system for physical and chemical studies of a variety of phenomena. The lipid assembling process that leads to vesicle formation is completely spontaneous; thus, liposomes are easy and fast to obtain.

The main advantages in the use of AMPs are that they are not toxic for eukaryotic cells, and that, with the membrane as their principal target, the raise of bacterial resistance against them seems unlikely. In the first part of this thesis, studies on the mechanism of membrane perturbation by AMPs will be presented.

The problem of AMPs selectivity will be also analyzed in detail. Selectivity is the capability of AMPs to distinguish between bacterial and mammalian cells, and to be toxic only for the former. The structural features that are essential for selectivity have not been clarified, yet; they will be investigated in Sect. 4.2 of Chap. 4 of this work.

The cell membranes represent a barrier also to the diffusion of drugs. In many cases, their internalization can be enhanced with the help of a carrier. Some peptides, called cell-penetrating peptides (CPPs), have the capability to cross-biological membranes, without damaging them, even when they are conjugated with a large cargo, i.e., constituted by a drug molecule. Section 4.3 of Chap. 4 will be dedicated to the possibility to modulate the biological activity of a CPP, to obtain a peptide with both penetrating and bactericidal functions.

The study of biological membranes is a fundamental topic of biology and biophysics. However, the particular characteristics of this system make it interesting also in different fields, like polymer science and nanoscience.

Due to its impermeability, the lipid membrane represents the envelope of a segregated environment. A chemical reaction can be carried out in this closed compartment, without occurring in the outer solution. In the second part of this thesis, liposomes will be exploited as nanotemplates for the photopolymerization of a hydrogel, allowing the formation of nanometric particles that can be used as drug delivery nanodevices for enzymatic therapies. The novelty of this approach resides in the fact that these nanoparticles have not been designed for the release of the enzyme in the organism, but to keep it within the gel, protecting it against inactivating factors, but allowing at the same time its catalytic function.

Contents

List of Publications Derived from this Thesis

- **Bobone, S.**, Miele, E., Cerroni, B., Nicolai, E., Di Venere, A., Placidi, E., Paradossi, G., Rosato, N., Stella, L. *Hydrogel nanoparticles for enzyme-based therapies*. Paper in preparation.
- **Bobone, S.**, Bocchinfuso, G., Park, Y., Palleschi, A., Hahm, K. S. and Stella, L. *The importance of being kinked: role of Pro residues in the selectivity of the helical antimicrobial peptide P5*. Journal of Peptide Science (2013) 19:758–769.
- **Bobone, S.**, van der Weert, M., Stella, L. *A reassessment of synchronous fluorescence in the separation of Trp and Tyr contribution in protein emission and in the determination of conformational changes*. Journal of Molecular Structure, in press (http://dx.doi.org/10.1016/j.molstruc.2014.01.004).
- **Bobone, S.**, Gerelli, Y., De Zotti, M., Bocchinfuso, G., Farrotti, A., Orioni, B., Palleschi, A., Sebastiani, F., Latter, E., Penfold, J., Senesi, R., Formaggio, F., Toniolo, C., Fragneto, G., Stella, L. *Membrane thickness and the mechanism of action of the short peptaibol trichogin GA IV*. Biochimica Biophysica Acta (2013) 1828:1013–1024.
- **Bobone, S.**, Roversi, D., Giordano, L., De Zotti, M., Formaggio, F., Toniolo. C., Park, Y., Stella. L. *Lipid dependence of antimicrobial peptide activity is an unreliable experimental test for different pore models*. Biochemistry (2012) 51:10124–10126.
- **Bobone, S.**, Piazzon, A., Orioni, B., Pedersen, J. Z., Nan, Y. H., Hahm, K. S., Shin, S. Y., Stella, L. *The thin line between cell-penetrating ant antimicrobial peptides: the case of Pep-1 and Pep-1-K*. Journal of Peptide Science (2011) 17:335–341.
- Bocchinfuso, G., **Bobone, S.**, Palleschi, A., Stella, L. *Fluorescence spectroscopy and molecular dynamics simulations in studies on the mechanism of membrane destabilization by antimicrobial peptides*. Cellular and Molecular Life Sciences (2011) 68:2281–2301.

Part I
Antimicrobial Peptides: Mechanism of Action, Selectivity and Biological Activity

Chapter 1
Introduction

1.1 Antimicrobial Peptides

Antimicrobial peptides (AMPs) are oligopeptides produced by the innate immune system of a large variety of organisms, as a first defense mechanism against bacteria. They were first discovered and isolated in the early 1980s, and nowadays about 1,000 different AMPs' sequences have been identified, in plants, animals, and bacteria [3, 16, 24, 33]. In humans, AMPs are mainly present on skin and in the oral and gastric mucous membranes [37].

AMPs exhibit a broad spectrum of antibacterial activity against both Gram (+) and Gram (−) microbes; some of them are also active against fungi, yeasts, and against biofilms generated by bacteria [7, 16]; more recently, an interesting anticancer activity of some AMPs has been observed [11, 23, 24, 25, 29]. AMPs are also investigated for the development of new antimicrobial strategies for plant protection [4]. In addition to their direct antimicrobial activity, AMPs play also a key role as regulators of the innate immune system and modulators of the of the immune response, being responsible for processes like chemokine production, angiogenesis, and acting as coadjuvants in promoting adaptive immunity [7].

AMPs are very diverse with respect to structural features such as length, charge, secondary structure, presence of disulfide bridges, etc. However, most AMPs are short, comprising 10–30 residues [24, 26, 38], and are characterized by a positive net charge, amphipathicity, and a helical structure when bound to the membrane, while they are very often unstructured in water.

The main target of AMPs is the bacterial cell membrane: they have the capability to perturb the lipid bilayer, altering its permeability and leading bacteria to death (Fig. 1.1).

S. Bobone, *Peptide and Protein Interaction with Membrane Systems*,
Springer Theses, DOI: 10.1007/978-3-319-06434-5_1,

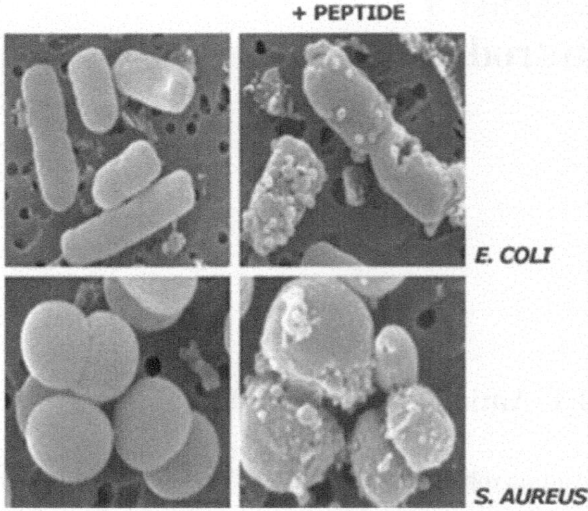

Fig. 1.1 Effects of AMPs on bacterial membranes: electron microscopy images of two different bacterial strains, before and after treatment with an AMP. The images show peptide-induced leakage of cell contents. Reproduced with permission from [28]

1.2 The Raising Problem of Multidrug Bacterial Resistance

The study of AMPs and of their characteristics is of great relevance because of the phenomenon of bacterial resistance. The discovery of penicillin and of the other antibiotic drugs marked a dramatic change in life expectancy and quality. However, since the first years of use of these "miracle drugs", bacterial resistance started to raise. The widespread use of antibiotics, not only for human therapy, but also in animal husbandry and agriculture, made this phenomenon even worse. In the last years, infections like tuberculosis or pneumonia became lethal again, bringing to an increasing number of deaths. In the past years multidrug resistant bacterial strains have been responsible for the so-called "hospital-acquired infections", but recently the contraction of drug-resistant infections became more common also outside the health centers. The Infectious Disease Society of America recently launched the 10×20 initiative, to support the development of ten new systemic antibacterial drugs, by 2020, trying to reverse the severe decline in the discovery of new antibacterial principles caused by the low profitability of these drugs for pharmaceutical companies [8].

In this scenario, it is easy to understand the importance of AMPs: due to their aspecific, "mechanic" effect on cell membranes, the occurrence of resistance is very unlikely. In fact, the mechanism of bacterial resistance often is given by alteration of the drug binding site or expulsion from the cell of the antibiotic drugs, or even by the modification of metabolic pathways. Thus, AMPs can bypass all these phenomena, because their binding to bacterial membranes is mainly driven by electrostatic interactions and by the hydrophobic effect.

1.3 Models for AMPs Activity

Several different models have been proposed in the literature to describe the mode of membrane permeabilization by AMPs. However, most AMPs have been found to act according to either one of the so called “barrel-stave” and “carpet” models (Fig. 1.2) [1, 2, 15, 20, 27, 34].

In the barrel-stave model, peptide molecules initially bind on the membrane surface, parallel to it. When the membrane-bound peptide concentration increases, reaching a certain threshold, peptides insert within the bilayer, forming transmembrane pores constituted by aggregated helices, in which the molecules interact laterally, and are arranged like the staves of a barrel. These structures are favored by the amphiphilic character of the peptide helices: in the aggregate, the hydrophilic side of the helix points towards the aqueous lumen of the pore, while the hydrophobic side interacts with the hydrophobic core of the membrane. The overall structure of the cylindrical, water-filled pore, resembles that of protein ion channels (Fig. 1.2). In this mechanism, the formation of the pore is a cooperative process: after the insertion of one peptide molecule inside the bilayer, other monomers are recruited to form the aggregates. Since the barrel-stave mechanism foresees peptide insertion in the hydrophobic core of the bilayer, and peptide aggregation, AMPs acting according to this model are usually highly hydrophobic and electrically neutral. One of the most studied pore-forming peptides acting according to this mechanism is alamethicin [1].

The carpet model is typical of cationic peptides in anionic bilayers: in this case, peptides interact strongly with the phospholipids polar headgroups by electrostatic attraction, covering the membrane surface like a carpet, causing a perturbation to the surface tension of the bilayer that leads to the formation of defects and eventually to membrane micellization. In contrast to the barrel-stave mechanism, the peptide interacts with the phospholipid headgroups throughout the entire process. In some cases it has been proposed a very specific structure for the peptide-generated membrane defects, which have been called “toroidal pores” [13, 19, 21]: in these structures, characterized by a toroidal shape, the lipid bilayer bends back onto itself and the two leaflets connect.

A physico-chemical interpretation of the mechanism of pore formation by AMPs, which can explain the driving forces for both the barrel-stave and the carpet models, is the two-state mechanism, proposed by Huang [12]. It suggests that, for low peptide/lipid (P/L) ratios, peptides bind to the membrane lying parallel to its surface. Above a certain P/L threshold value (P/L*), which depends both on the specific peptide and on the lipid composition of the bilayer, peptides attain a transmembrane helical orientation. The surface-parallel orientation is called S state, while the transmembrane, inserted orientation is called I state. The I state is associated with the formation of both toroidal pores or barrel-stave pores, while the S state is considered “functionally inactive”.

When peptides bind to the membrane surface, they cause a stretching of the membrane area. This distortion has a cost in terms of energy. Thus, the free energy

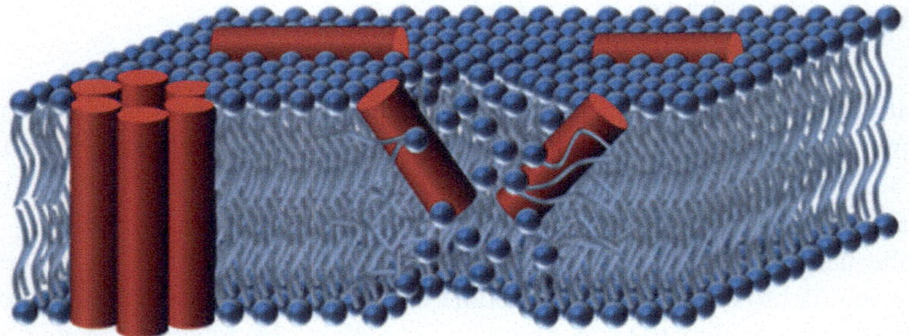

Fig. 1.2 Representation of barrel-stave pores formed by AMPs (*left side*) and of carpet model (*right side*)

of adsorption on the membrane surface is composed by a negative term of binding energy and a positive term of bilayer deformation energy. The bilayer deformation is proportional to P/L and the energy cost to $(P/L)^2$, according to the elasticity theory. Thus, the energy of the S state increases with the $(P/L)^2$ term. At low concentration, the energy level of S state is lower than the one of the I state, but at sufficiently high concentrations the energy for S state exceeds the energy of the I state, and the S-to-I transition occurs [12].

These models can provide an explanation to the bactericidal activity of some AMPs, but they cannot be easily applied to all of them. For many AMPs s the mechanism of pore formation has not been established. In Sect. 2.4 of Chap. 2 this goal has been reached for the peptide trichogin GAIV. This is an example of a widely studied, well-characterized peptide, whose mechanism is still debated, despite the large number of available data. Several studies have been carried out to characterize its pore-forming activity, with apparently ambiguous results. Using a combination of spectroscopic techniques, and molecular dynamics simulations, the behavior of this peptide has been completely clarified and the apparent contradictions have been solved.

1.4 Selectivity

Selectivity is one of the most fundamental qualities of AMPs, in view of their clinical application. Therefore a thorough understanding of the features determining this attribute is essential. Since the cell membrane is the main target of AMPs, selectivity must be dictated by differences between the membrane properties of bacteria and eukaryotes, and indeed these bilayers exhibit three relevant differences: the bacterial membrane contains a high fraction of negatively charged lipids (such as such as phosphatidylglycerol or cardiolipin), while eukaryotic bilayers don't; transmembrane potential is much more negative inside in bacterial than in eukaryotic membranes; bacterial membranes are essentially free of sterols, while eukaryotic bilayers contain a high fraction of cholesterol [18, 35] (Tables 1.1 and 1.2).

Table 1.1 Lipid composition of bacterial membranes. *OM* outer membrane, *CM* cytoplasmic membrane, *PG* phosphatidyl glycerol, *CL* cardiolipin, *PE* phosphatidyl ethanolamine [14]

Bacterial strain		PG	CL	PE
Gram-negative				
Eschericha coli	OM	3	6	91
	CM	6	12	82
Salmonella typhimurium	OM	17	2	81
	CM	33	7	60
Pseudomonas cepacia	OM	13	0	87
	CM	18	0	82
Gram-positive				
Bacillus megaterium	CM	40	5	40
Bacillus subtilis	CM	29	47	10
Micrococcus luteus	CM	26	67	0
Staphylococcus aureus	CM	57	5	0

Table 1.2 Lipid composition of eukaryotic red blood cells. *PC* phosphatidyl choline, *SM* sphingomyelin, *PS* phosphatidyl serine [14]

Organism	PC	SM	PE	PS
Rabbit	33.9	19.0	31.9	12.2
Man	34.7	20.1	28.0	14.3
Pig	23.3	26.5	29.7	17.8
Sheep	-	51.0	26.2	14.1

NMR studies have demonstrated that the inclusion of cholesterol within lipid membranes causes a reduction in the disrupting activity of a lytic peptide [17]. Ergosterol has the same effect but to a minor extent. Cholesterol usually causes an increment in the lipid chain order [32], and this might explain why it inhibits the membrane-perturbing activity of AMPs.

A different interpretation of similar results, obtained with different peptides, has also been proposed [30], indicating that the presence of sterols reduces the possibility for the peptide to intercalate within the bilayer, and consequently the fraction of membrane-bound peptide in the water-partition equilibrium is lower. The phenomenon has been interpreted using the two state model: in the presence of sterols, the threshold P/L* value increases.

An increase in peptide charge causes an enhanced binding to anionic membranes and therefore improves antibacterial activity [18, 35]. Since binding to neutral membranes and toxicity are not significantly affected by this parameter, increasing peptide charge usually leads also to a higher selectivity. However, several studies have shown that charge can increase peptide activity only up to a certain point, because after it has reached a value of about $+10\ e$, further increases do not lead to any advantage in peptide properties [18, 35, 36].

Hydrophobicity, on the other hand, contributes to peptide affinity for both charged and neutral membranes. It has been shown that, by increasing peptide hydrophobicity by keeping all other properties constant, two thresholds are observed [6, 10, 22, 31]: the first, at lower hydrophobicity values, must be reached to have antimicrobial activity. The second threshold is at significantly higher hydrophobicity values, and must be reached to observe toxicity. The difference between the two hydrophobicity values is due to the fact that electrostatic interactions contribute to peptide binding to bacterial membranes but not to those of the host organism. In some cases, at very high hydrophobicity values, loss of activity is observed [6], probably due to aggregation [9]. Therefore, an optimal range of hydrophobicity values exists, in which peptides exhibit antimicrobial activity, but no significant toxicity.

Helicity is an important structural characteristic for membrane association, and it also plays a role in selectivity: the introduction of amino acids that promote the helix formation, or removal of helix-breaking amino acids, usually have the effect to increase the antimicrobial activity, but also to dramatically enhance peptide's hemolytic and cytotoxic activity, with the result of a loss of selectivity. An increased peptide helicity, as well as hydrophobic moment, is responsible for higher hydrophobic interactions, and thus, for an enhanced binding and lytic activity to neutral or weakly charged membranes.

Notwithstanding all these considerations, the design of AMPs with enhanced activity, which at the same time maintain a high selectivity, is still a challenge. In particular, the role of AMPs structure in determining their selectivity is still unclear. For this reason a detailed study of all the structural characteristics that play a role in selectivity is fundamental. In Sect. 4.2 of Chap. 4 of this thesis, the role of an important feature, will be illustrated, i.e. the presence of a proline residue in the center of the amino acidic sequence, that is common to several selective peptides.

1.5 Relationship Between Function and Structure

The structural characteristics of AMPs play a very important role in their selectivity and biological activity. It is possible to modulate the biological activity of a peptide by operating only a few substitutions into the amino acidic sequence. This is true not only for AMPs, but also for other classes of membrane-active peptides, one of which is represented by cell-penetrating peptides (CPPs). These molecules are able to translocate across biological membranes and carry large cargoes inside cells, without damaging the bilayer. In many cases, CPPs and AMPs share some common characteristics, like positive net charge, amphipathicity, and a helical structure after membrane association. Thus, it is possible to hypothesize that a CPP can be turned into an AMP with a small number of substitutions in its amino acidic sequence.

In Sect. 2.4, it will be shown that a CPP, through a few amino acid substitution can become an AMP, with a strong antibacterial activity, without losing its ability to translocate across lipid bilayers.

References

1. Baumann G, Mueller P (1974) A molecular model of membrane excitability. J Supramol Struct 2:538–557
2. Bechinger B (1997) Structure and function of channel-forming peptides: magainins, cecropins, melittin and alamethicin. J Membr Biol 156:197–211
3. Brogden KA (2005) Antimicrobial peptides: pore formers or metabolic inhibitors in bacteria? Nature 3:238–250
4. Butu M, Butu A (2011) Antimicrob Peptides-Nat Antibiot Rom Biotech Lett 16:6135–6145
5. Cafiso DS (1994) Alamethicin: a peptide model for voltage gating and protein-membrane interactions. Ann Rev Biophys Biomol Struct 23:141–165
6. Chen Y, Guarnieri MT, Vasil AI, Vasil ML, Mant CT, Hodges RS (2007) Role of peptide hydrophobicity in the mechanism of action of -helical antimicrobial peptides. Antimicrob Agents Chemother 51:1398–1406
7. Fjell CD, Jan A, Hiss JA, Hancock REW., Schneider G (2012) Designing antimicrobial peptides: form follows function. Nat Rev Drug Discov 11:37–51
8. Garwood J (2011) Antibiotic resistance. Who will win the fight? Lab Times 7:19–24
9. Gatto E, Mazzuca C, Stella L, Venanzi M, Toniolo C, Pispisa B (2006) Effect of peptide lipidation on membrane perturbing activity: a comparative study on two trichogin analogues. J Phys Chem B 110:22813–22818
10. Glukhov E, Burrows LL, Deber CM (2008) Membrane interactions of designed cationic antimicrobial peptides: the two thresholds. Biopolymers 89:360–371
11. Hoskin DW, Ramamoorthy A (2008) Studies on anticancer activity of antimicrobial peptides. Biochim Biophys Acta 1778:357–375
12. Huang HW (2000) Action of antimicrobial peptides: two state model. Biochemistry 39:8347–8352
13. Lohner K, Blondelle SE (2005) Molecular mechanisms of membrane perturbation by antimicrobial peptides and the use of biophysical studies in the design of novel peptide antibiotics. Comb Chem High Throughput Screen 8:241–256
14. Lohner K, Sevcsik E, Pabst G (2008) Liposome-based biomembranes mimetic systems: implication for lipid-peptide interactions. In: Leitmannova-Liu A (ed) Advances in planar lipid bilayers and liposomes, vol 6. Elsevier, Amsterdam, pp 103–137
15. Ludtke SJ, He K, Heller WT, Harroun TA, Yang L, Huang HW (1996) Membrane pores induced by magainin. Biochemistry 35:13723–13728

16. Mahalka AK, Kinnunen PK (2009) Binding of amphipathic alpha-helical antimicrobial peptides to lipid membranes: lessons from temporins B and L. Biochim Biophys Acta 1788:1600–1609
17. Mason AJ, Marquette A, Bechinger B (2007) Zwitterionic phospholipids and sterols modulate antimicrobial peptide-induced membrane destabilization. Biophys J 93:4289–4299
18. Matsuzaki K (2009) Control of cell selectivity of antimicrobial peptides. Biochim Biophys Acta 1788:1687–1692
19. Matsuzaki K (2001) Molecular mechanism of membrane perturbation by antimicrobial peptides. In: Lohner K (ed) Development of novel antimicrobial agents: emerging strategies, Horizon Scientific Press, UK, pp 167–181
20. Matsuzaki K, Murase O, Fuiji N, Miyajima K (1996) An antimicrobial peptide, magainin-2, induced rapid-flip flop of phospholipids coupled with pore formation and peptide translocation. Biochemistry 35:11361–11368
21. Oren Z, Shai Y (1998) Mode of action of linear amphipathic alpha-helical antimicrobial peptides. Biopolymers 47:451–463
22. Papo N, Oren Z, Pag U, Sahl HS, Shai Y (2002) The consequence of sequence alteration of an amphipathic α-helical antimicrobial peptide and its diastereomers. J Biol Chem 277:33913–33921
23. Papo N, Shai Y (2005) Host defense peptides as new weapons in cancer treatment. Cell Mol Life Sci 62:784–790
24. Reddy KV, Yedery RD, Aranha C (2004) Antimicrobial peptides: premises and promises. Int J Antimic Agents 24:536–547
25. Schweizer F (2009) Cationic amphiphilic peptides with cancer-selective toxicity. Eur J Pharmacol 625:190–194
26. Shai Y (1999) Mechanism of the binding, insertion and destabilization of phospholipid bilayer membranes by alpha-helical antimicrobial and cell non-selective membrane-lytic pepties. Biochim Biophys Acta 1462:55–70
27. Shai Y (2002) Mode of action of membrane active antimicrobial peptides. Biolpolymers (Pept Sci) 66:236–248
28. Skerlavaj B, Benincasa M, Risso A, Zanetti M, Gennaro R (1999) SMAP-29: a potent antibacterial and antifungal peptide from sheep leukocytes. FEBS Lett 463:58–62
29. Slaninová J, Mlsová V, Kroupová H, Alán L, Tumová T, Monincová L, Lenka Borovicková L, Vladimír Fucík V, Cevrovsky V (2012) Toxicity study of antimicrobial peptides from wild bee venom and their analogs toward mammalian normal and cancer cells. Peptides 33: 18–26
30. Sood R, Kinnunen PKJ (2008) Cholesterol, lanosterol, and ergosterol attenuate the membrane association of LL-37(W27F) and temporin L. Biochim Biophys Acta 1778:1460–1466
31. Stark M, Liu LP, Deber CM (2002) Cationic Hydrophobic Peptides with Antimicrobial Activity. Antimicrob Agents Chemother 46:3585–3590
32. Urbina JA, Pekerar S, Le HB, Patterson J, Montez B, Oldifeld O (1995) Molecular order and dynamics of phosphatidylcholine bilayer membranes in the presence of cholesterol, ergosterol and lanosterol: a comparative study using 2H-, 13C- and 31P-NMR spectroscopy. Biochim Biophys Acta 1238:163–176
33. Wimley WC (2010) Describing the mechanism of antimicrobial peptide action with the interfacial activity model. ACS Chem Biol 5:905–917
34. Yang LY, Harroun TA, Weiss TM, Ding L, Huang HW (1991) Barrel-stave or toroidal model? a case study on melittin pores. Biophys J 81:1475–1485
35. Yeaman MR, Yount NY (2003) Mechanisms of antimicrobial peptide action and resistance. Pharmacol Rev 55:27–55
36. Zelezetsky I, Tossi A (2006) Alpha-helical antimicrobial peptides-using a sequence template to guide structure-activity relationship studies. Biochim Biophys Acta 1758:1436–1449
37. Zhang L, Fjalla TJ (2010) Potential therapeutic application of host defense peptides. Methods Mol Biol 618:303–327
38. Zhang L, Rozek A, Hancock REW (2001) Interaction of cationic antimicrobial peptides with lipid membranes. J Biol Chem 276:35714–35722

Chapter 2
Techniques

In this chapter the rudiments of the two main experimental techniques exploited in this work, i.e. fluorescence spectroscopy and neutron reflectivity, will be briefly discussed.

2.1 Fluorescence Spectroscopy

Molecular fluorescence is the emission of light from a molecule—the fluorophore—after being excited by the absorption of a photon. Fluorescence spectroscopy is a powerful technique, which provides information on the structure and dynamics of biomolecules, and for this reason it is one of the most widely used methodologies in biophysical studies. Proteins and peptides are often endowed with intrinsic fluorescence, due to the presence of Trp or Tyr residues. A great advantage of this technique is its sensitivity, which gives the possibility to work at low concentration of fluorophores.

A fluorescent molecule can be excited by light absorption from the ground state (S_0) to a higher electronic state (S_1 or S_2 in Fig. 2.1) using appropriate wavelengths (λ_1 or λ_2) [9], corresponding to the maxima of absorption. In each electronic state, the fluorophore can occupy different vibrational energy levels, but it usually rapidly relaxes to the lower vibrational level of S_1, transferring the excess of its vibrational energy to the solvent by molecular collisions. If the fluorophore is excited to the second singlet state S_2, it rapidly decays to S_1. This process is called internal conversion.

From this level, the molecule spontaneously returns to the ground state. The relaxation to the ground state is in this case a spin-allowed transition, and occurs in 10^{-5}–10^{-8} s, with the emission of a photon. If an intersystem crossing takes place, leading to a triplet excited state, the phenomenon of phosphorescence occurs.

The radiative decay is in competition with all other nonradiative relaxation pathways; a quantitative way to characterize the importance of the radiative

S. Bobone, *Peptide and Protein Interaction with Membrane Systems*,
Springer Theses, DOI: 10.1007/978-3-319-06434-5_2,

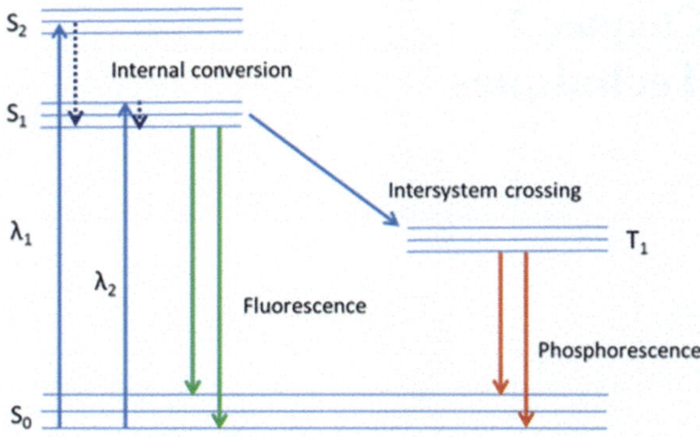

Fig. 2.1 Schematic representation of the phenomenon of fluorescence with the Jablonski diagram

process with respect to the others is the quantum yield, defined as the ratio between emitted and absorbed photons:

$$q = \frac{n_f}{n_a} = \frac{k_r}{k_r + k_{nr}}$$

in which k_r is the rate constant of radiative processes, while k_{nr} is the rate constant of nonradiative processes. Therefore, this parameter represents the probability that an excited molecule returns to the ground sate by emitting a photon. It can be close to unit if k_{nr} is much smaller than k_r [7].

2.1.1 Steady-State Fluorescence

In steady-state measurements, the sample is illuminated with a continuous light beam and the intensity of emission is recorded as a function of wavelength. The steady-state condition is reached in few ns; thus, a constant fraction of fluorophores is excited during the collection of the emission spectrum [7]. This spectrum represents the probability distribution of all the transitions from the lower vibrational state of S_1 to each vibrational state of S_0. The emission spectrum is characteristic for each fluorophore; the shape and position of spectral peaks are extremely sensitive to the molecular properties of the medium. This is due to the perturbation of fluorophore's energy levels caused by electrostatic or dipole-dipole interactions, which are stronger in polar solvents. The effects of solvent polarity are the main origin of the difference between absorption and emission wavelengths (the Stoke's shift) [1, 7].

Most of the peptides investigated in this work have one or more Trp residues in their sequence, and therefore are intrinsically fluorescent. The intensity and the position of the maximum in the emission spectrum can give information about the position of a peptide in the lipid membrane. When the peptide is inserted within the membrane, its intensity usually increases, because the more viscous nature of the lipid bilayer reduces the rate of nonradiative decays with respect to the aqueous phase. Furthermore, the position of the maximum emission is shifted to lower wavelengths, because of the lower polarity of the membrane environment.

A measurement of the blue shift that occurs when a peptide binds to a lipid bilayer can be provided by the average wavelength of the spectrum, which is defined as [1]

$$\langle \lambda \rangle = \Sigma_i \lambda_i I_i / \Sigma_i I_i$$

where λ_i represents each wavelength of the emission spectrum and I_i is the corresponding intensity. Therefore, the variation of emission intensity and of the wavelength of maximum emission (or of the average wavelength) represents an evidence of peptide–membrane association.

2.1.2 Time-Resolved Fluorescence

Time-resolved fluorescence provides a characterization of the excited state lifetime. When a homogeneous population of fluorescent molecules is excited with a short light pulse, the decay of the emitted intensity can be described by a monoexponential law:

$$F(t) = F_0 \exp(-t/\tau)$$

where F_0 is the fluorescence intensity immediately after the pulse and τ is the fluorescence lifetime, i.e. The average time that a molecule spends in the excited state before the relaxation to the ground state. τ is related to the rate constants of radiative and nonradiative processes by the following equation:

$$\tau = \frac{1}{k_r + k_{nr}}$$

The steady-state emission intensity is related to the fluorescence lifetime by the following equation:

$$I_{ss} = \int_0^\infty F_0 e^{-t/\tau} dt = F_0 \tau$$

Time-resolved fluorescence is particularly useful to characterize heterogeneous systems in which the fluorophore exists in different states (different conformations, aggregation states, phases). In this case, each population of molecules is characterized by a different lifetime, due to the fact that their k_{nr} values are different (while k_r is quite similar, being independent from the environment of the fluorophore). The lifetime decay is given by the contribution of all the present species:

$$F(t) = F(0)\sum_i \alpha_i e^{-t/\tau i}$$

Here the pre-exponential parameters α_i correspond to the molar fractions of the different species present in the sample.

A further advantage of lifetime measurements is that they are much less affected by experimental artifacts than steady-state experiments. For instance, they are independent on fluorophore concentration and do not suffer from inner-filter effect (apparent reduction in steady-state intensity due to absorption of the exciting or emitted light by the sample itself).

2.1.3 FRET

The phenomenon called Förster Resonance Energy Transfer (FRET) is a very useful tool to determine intermolecular or intramolecular distances. This process consists in the transfer of energy from a donor, in its excited state, to an acceptor, in its ground state [7]. It is a resonance process due to the dipolar interaction between the two fluorophores, and it does not involve the emission of photons. The necessary condition for FRET to occur is that the donor's emission spectrum overlaps, at least partially, to the acceptor's absorption. The time constant of the FRET phenomenon is described by the following equation

$$k_T = \frac{1}{\tau_D} \cdot \frac{3\kappa^2}{2} \cdot \left(\frac{R_0}{r}\right)^6$$

where τ_D is the donor's lifetime in the absence of acceptors, κ^2 is a factor describing the relative orientation of the transition dipoles of the two molecules, assumed equal to 2/3 if they are moving rapidly during the fluorescence lifetime; r is the distance between the two fluorophores; R_0 is called the Förster radius, that is, the distance occurring between donor and acceptor when the transfer efficiency is 0.5. Förster distances are typically in a range between 15 and 60 Å, which is a very interesting range, since it includes for example the typical size of proteins, or the thickness of a lipid bilayer. From an experimental point of view, the observed phenomenon is a decrease in the fluorescence intensity in the donor's spectrum, due to the decrease in its fluorescence lifetime and in its quantum yield, and an increase in that of the acceptor. Indeed, the rate constant of the excited state decay of the donor is

$$k^D = k^D_{nr} + k^D_r + k_T$$

where k_r^D and k_{nr}^D are the rate constants of radiative and of all other nonradiative processes of the unperturbed donor and k_T is the rate constant of the energy transfer process.

FRET efficiency is defined as the probability that an excited donor relaxes to the ground state by energy transfer

$$E = \frac{k_T}{k_T + k_r^D + k_{nr}^D}$$

It is strictly related to the distance between donor and acceptor molecules, as shown by the following equation

$$E = \frac{1}{1 + \left(\frac{3}{2}k^2\right)^{-1}\left(\frac{r}{R_0}\right)^6}$$

This result indicates that FRET can be exploited to investigate the relative position of two fluorophores in a biological environment, and also to explore interactions occurring in the range of several Ångstroms. The range of distances in which FRET can be detected is comprised approximately between $R_0/2$ and $2R_0$.

In our case, FRET experiments are suitable to establish the interaction of a peptide with the lipid bilayer, exploiting the energy transfer occurring between the peptide Trp residues and a membrane-bound acceptor [1].

2.1.4 Fluorescence Anisotropy

Fluorescence anisotropy is used to determine the rotational dynamics of a fluorophore. Since molecular motions are significantly slower inside a lipid bilayer than in water, this technique can be used also to assess peptide-membrane association.

In a fluorescence anisotropy experiment the sample is excited with polarized light. The molecules in the sample are randomly oriented before the excitation, but the probability for a molecule to be excited is dependent on the factor $\cos^2\theta$, where θ is the angle between the directions of the polarization of the exciting light and of the transition dipole of the fluorophore. Thus, the molecules having the transition dipole oriented in the same direction of the exciting light have the highest probability of excitation. This phenomenon is called photoselection. The light emitted from the photoselected fluorophores is also polarized preferentially along the direction of the emission transition dipole, again according to a $\cos^2\theta$ law. By measuring the degree of polarization of the emitted light it is therefore possible to assess the degree of molecular motions that take place during the fluorophore's excited state lifetime.

If the molecular motions are very fast with respect to τ, the preferential orientation is lost during the lifetime decay, and the emitted light is unpolarized, due to the random orientation of the transition dipoles. By contrast, if the motions are slow, the emitted light is anisotropic (i.e. exhibits a preferential polarization). Thus, anisotropy measurements can provide information about the rotational motions of a fluorophore in the nanoseconds time scale, and about the viscosity of its environment.

Experimentally, the anisotropy is measured as

$$r(t) = \frac{I_{vv}(t) - GI_{vh}(t)}{I_{vv}(t) + 2GI_{vh}(t)}$$

where the subscripted indexes indicate the vertical (v) or horizontal (h) orientation of excitation and emission polarizers with respect to the instrumental plane, and G is an instrumental parameter, defined as

$$G = I_{hv}/I_{hh}$$

2.2 Neutron Reflectivity

Neutron spectroscopy techniques are a powerful tool for the study and characterization of biological systems and soft matter, like lipid membranes, rich in light atoms C, H, N and O, whose nuclei are good neutron scatterers. [3, 4, 6].

Since neutrons interact with atomic nuclei only weakly, they can penetrate deeply within samples, and provide information about complex structures and buried interfaces [3].

Neutron reflectivity allows the characterization of the composition and of the irregularities of biological materials, in a length scale between 10 and 5000 Å.

In the specular neutron reflectivity technique, the reflected intensity is measured as a function of the incident angle [2]. Experiments are performed at a grazing angle, usually varying from 0.01 to 5° [3]; sample and detector are moved in order to keep the incident and reflected angles equal.

Specular reflectivity is defined as the ratio between the reflected and the incoming beam intensities, and it is measured as a function of the scattering vector, Q, perpendicular to the surface plane:

$$R = \frac{16\pi^2}{Q^2} \left|\widetilde{\rho}(Q)^2\right|$$

where Q, is defined as

$$Q = \frac{4\pi}{\lambda} sen\, \theta$$

and $\widetilde{\rho}(Q)$ is the Fourier transform of $\rho(z)$, the scattering length density (SLD), that is the parameter providing informations about the structure along the normal axis to the surface. Thus, there is a direct relationship between reflectivity and structure. The SLD is formally defined as $\rho(z) = \Sigma_i v_i b_i$, where v_i is the number of nuclei of type i per volume unit and b_i is the neutron scattering length of those nuclei. During the experiments, a profile of the reflectivity R as a function of Q is acquired.

An important characteristic of neutron spectroscopy is its extreme and unique sensitivity to the substitution of hydrogen by deuterium, due to the fact that neutrons are scattered very differently from these two nuclei. Indeed, the scattering lengths values are –0.37 for hydrogen and 0.67 for deuterium [5]. Thus, it is possible to perform experiments in different H_2O/D_2O mixtures, to enhance the signal of some atoms and minimize that of other ones. This technique is known as the *contrast variation method*; all the profiles obtained in different media are simultaneously fitted. This method is used to have enough constrains to exactly define the physically meaningful model.

The data fitting procedure is based on the Parratt's recursion relation [8], and it often involves the construction of a model of the interface that can be represented by a series of parallel layers of homogeneous material. In the present case, the lipid bilayer has been divided into three sublayers, i.e. the outer headgroups region, the tails region, and the inner headgroups region.

References

1. Bocchinfuso G, Bobone S Palleschi A, Stella L (2011) Fluorescence spectroscopy and molecular dynamics simulations in studies on the mechanism of membrane destabilization by antimicrobial peptides. Cell Mol Life Sci 68:2281–2301
2. Dabowska AP, Fragneto G, Hughes AV, Quinn PJ, Lawrence MJ (2009) Specular neutron reflectivity studies on the interaction of Cytochrome c with supported phosphatidylcholine bilayers doped with phosphotidylserine. Langmuir 25:4203–4210
3. Fragneto, G. and Rheinstaedter (2007) Structural and dynamical studies from bio-mimetic systems: an overview. C R Phys 8:865–883
4. Fragneto G, Graner F, Charitat T, Dubos P, Bellet-Amalric EB (2000) Interaction of the third helix of antennapedia homeodomain with a deposited phospholipid bilayer: a neutron reflectivity structural study. Langmuir 16:4581–4588
5. Jacrot B (1976) The study of biological structures by neutron scattering from solution. Rep Prog Phys 39:911
6. Krueger S (2001) Neutron reflection from interfaces with biological and biomimetic materials. Curr Op in Coll Interface Sci 6:111–117
7. Lakowicz JR (2006) Principles of Fluorescence Spectroscopy, 3rd edn. Springer, berlin
8. Parrat LG (1954) Surface studies of solids by total reflection of X-rays. Phys Rev 95:359–369
9. Skoog DA, Leary JJ (1992) Principles of instrumental analysis. Saunders College Pub, New York

Chapter 3
Materials and Methods

3.1 Materials

All phospholipids were purchased from Avanti Polar Lipids (Alabaster, AL, USA).

Spectroscopic grade methanol, ethanol and chloroform were purchased from Carlo Erba Reagenti (Milano, Italy). Carboxyfluorescein (CF), Triton-X 100, Sephadex-G50 and 1,6-Diphenyl-1,3,5-hexatriene (DPH) were purchased from a Sigma Aldrich (Germany).

The fluorophore sodium-binding benzofuran isophthalate (SBFI) was purchased from Invitrogen (Eugene, USA).

3.2 Peptide Synthesis

All peptides used for this study were prepared by the standard Fmoc-based solid-phase method [14]. Pep-1, Pep-1-K, P5 and P5 analogues were synthesized by the Research Center for Proteinous Materials (Chosun University, Korea). Trichogin GA IV and alamethicin were synthesized in the laboratory of prof. C. Toniolo, University of Padova. The peptide purity was checked by MALDI-TOF MS (matrix-assisted laser-desorption ionization-time-of-flight mass spectrometry) (Shimadzu, Japan) in the former case and by Electrospray ionization (ESI)-MS, performed on a PerSeptive Biosystem Mariner instrument (Framingham, MA), in the latter case.

Peptides used for all the experiments were dissolved in spectroscopic-grade methanol, except for Pep-1-K, which was solubilized in phosphate buffer 10 mM at pH 7.4.

All the concentrations of Pep-1-K and P5 analogues stock solutions were determined by measuring Trp absorbance at 280 nm. Trichogin GA IV and alamethicin concentrations were determined by measuring the absorbance of peptide bonds at 214 nm [12].

S. Bobone, *Peptide and Protein Interaction with Membrane Systems*,
Springer Theses, DOI: 10.1007/978-3-319-06434-5_3,

3.3 Liposomes Preparation

Phospholipid vesicles composed of egg yolk phosphatidylcholine (ePC)/phosphatidylglycerol (ePG) (2:1 molar ratio), ePC/cholesterol (1:1 molar ratio), 1,2-Dimyristoyl-sn-glycero-3-phosphocholine (DMPC), or 1-palmitoyl-2-oleoyl-sn-glycero-3-phosphocholine (POPC) were prepared with the lipid film method [20], by dissolving lipids in a methanol-chloroform 1:1 mixture, in a lipid amount such to obtain a 10 mM stock solution. The solvent was then gently evaporated in a rotary vacuum system, until a homogeneous, thin film was formed. Complete solvent evaporation was ensured by applying a rotary vacuum pump for at least 2 h. Successively, the film was hydrated with a physiological buffer (phosphate buffer 10 mM, NaCl 140 mM, pH 7.4). After 10 freeze-thaw cycles, liposomes were extruded with a LiposoFast Extruder (Avestin GMBH, Mannheim, Germany) through two stacked polycarbonate membranes with pores 50, 100 or 200 nm in diameter. The final lipid concentration was estimated using Stewart's phospholipid assay [22].

For DPH fluorescence anisotropy experiments, liposomes were prepared adding 1 % DPH to the lipid mixture from a stock methanolic solution before film formation.

For depth-dependent quenching experiments, nitroxide-labeled liposomes were produced by adding the labeled lipids to the initial chloroform solution (7 % molar fraction). The final mixture was ePC/ePG/doxyl-PC in 60:33:7 molar ratios. Spin label content was controlled directly on the final liposomes by double integration of the EPR spectra of an aliquot of the liposomes dissolved in isopropanol [3]. All liposome preparations contained the same amount of spin labels, within a 10 % error.

For ion-leakage experiments, ePC/ePG lipids (2:1 molar ratio) were hydrated with a solution containing Tris buffer (10 mM), pH 7.4, KCl (150 mM), and the SBFI probe (0.5 mM), and extruded through pores of 200 nm diameter. The vesicles were then eluted on a Sephadex G-50 column, in the same buffer, to remove the unencapsulated probe. Finally, liposomes were diluted (to a 30 μM lipid concentration) in a Na^+ containing buffer (10 mM Tris, pH 7.4, 150 mM NaCl).

For CF leakage experiments, the lipid film was obtained as already described, and then hydrated with a 30 mM CF solution, prepared in 10 mM phosphate buffer, pH 7.4, and titrated with 85 mM NaOH to make CF water-soluble, containing 80 mM NaCl to make it isotonic to the dilution buffer (phosphate buffer 10 mM, NaCl 140 mM, pH 7.4, 270 mOsm). Liposomes were then eluted on a 40 cm Sephadex G-50 column to remove all the unentrapped dye.

The percentage of leakage induced by the peptide was determined as

$$\%\ \text{Leakage} = \frac{F - F_0}{F_{100-F_0}}$$

where F is the fluorescence signal recorded 20 or 15 minutes after peptide addition, F_0 is the initial fluorescence of the vesicles, and F_{100} is the fluorescence obtained adding a detergent, Triton-X, to the vesicles solution, to allow their complete solubilization.

Giant unilamellar vesicles (GUVs) were prepared by the electroformation method [1, 21]. The lipid film was obtained by the spin-coating technique, using 250 μl of a 5 mM lipid solution in a chloroform/acetonitrile (95:5) mixture. The solution was spread on the charged side of an ITO-coated glass slide, using a speed of 600 rpm for 5 min. The solvent was then completely removed under vacuum for at least 1 h. After lipid film formation, the electroformation chamber was assembled using a second ITO-coated glass slide and a 1.5 mm thick silicon spacer (Fig. 3.1). The chamber was then filled with a 0.30 M sucrose solution, and a 1.5 V (peak to peak), 10 Hz potential was applied for 1 h, and then switched to 4 V, 4 Hz for 15 min to favor detachment of GUVs from the electrode. The solution contained in the electroformation chamber was gently removed, and diluted 300 times in buffer (pH 7.4, 10 mM phosphate, 140 mM NaCl, 0.1 mM EDTA). The lipid composition of GUVs was 66 % ePC, 33 % ePG and 1 % rhodamine-labeled phosphatidylethanolamine (Rho-PE). Glass slides employed during fluorescence imaging were previously treated with Sigmacote (Sigma Aldrich). The FITC-labeled peptide Pep-1-K (0.7 μM) and the lipid membrane could be observed independently by imaging the green fluorescein emission and the red Rho-PE fluorescence in a Nikon Ti Eclipse confocal laser scanning microscope (Nikon, Tokyo, Japan).

3.4 Spectroscopic Measurements

UV-visible spectra were acquired using a Cary 100 Scan UV-visible spectrophotometer (Varian). Fluorescence spectra were collected with a Fluoromax-2 spectrofluorimeter (Horiba). The temperature of the sample was controlled within 0.1 °C with a thermostatted cuvette holder. CD spectra were collected using a Jasco J 600 spectropolarimeter in the wavelength range between 195 and 250 nm. Experiments were performed in 0.1 cm pathlength cells, in phosphate buffer (10 mM, pH 7.4), in 30 mM SDS, in 50 % TFE and in the presence of ePC/ePG (2:1) liposomes with a nominal diameter of 50 nm at a lipid concentration of 250 μM. Peptide concentration was between 10 and 30 μM.

3.5 Turbidity Measurements

Peptide-induced vesicle aggregation was followed by measuring sample turbidity at 400 nm, after adding increasing lipid amounts to a 1 μM peptide solution for all investigated peptides (100 nm vesicles).

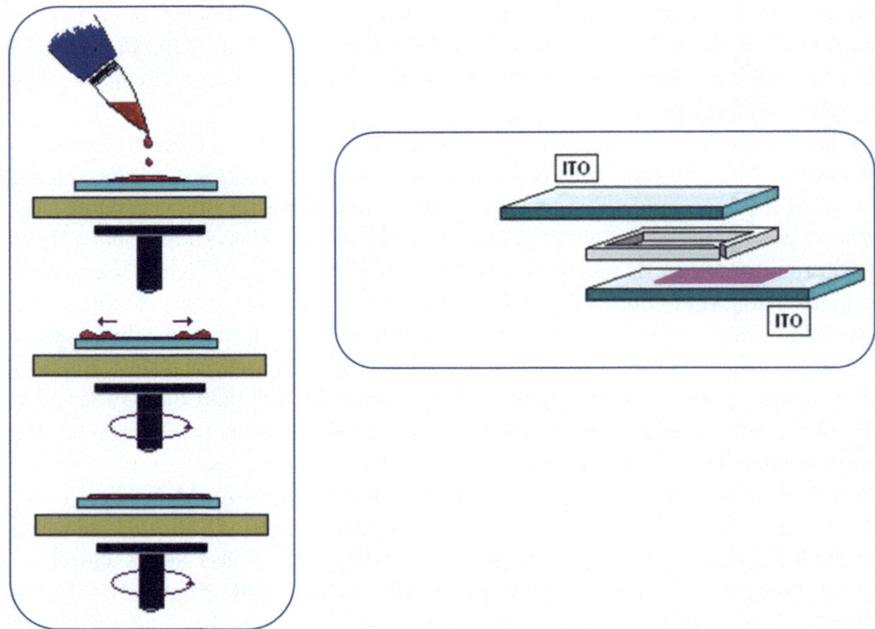

Fig. 3.1 Schematic representation of the spin-coating process (*left panel*) and of the GUVs formation chamber (*right panel*)

3.6 Osmolarity Measurements

The osmolarity of the liposomes hydration and dilution solutions were checked with a freezing-point depression FISKE 210 osmometer (Advanced Instruments, Norwood, USA). The calibration was carried out using 50 and 850 mOsm/Kg standard solutions.

3.7 Membrane-Perturbing Activity Experiments

For CF-leakage experiments with peptides P5 and P5Del, 0.03 μM peptide was added to a solution of 200 nm CF loaded-liposomes, at different peptide to lipid ratios. The CF fluorescence signal of liposomes was followed at $\lambda_{exc.} = 490$ nm and $\lambda_{em.} = 520$ nm. Total leakage was obtained by adding 1 mM Triton X-100. The fraction of peptide-induced leakage was determined 15 min after peptide addition to the liposomes solution, by the increase in the fluorescence signal [13, 20]. For trichogin GA IV and alamethicin, the same experiments were performed by adding the peptide to a 200 μM solution of 200 nm CF loaded-liposomes.

Peptide effects on membrane order and dynamics were determined by measuring the fluorescence anisotropy of 100 nm DPH labeled-DMPC liposomes, at a $\lambda_{exc.} = 350$ nm and $\lambda_{em.} = 450$ nm, at different temperatures or at different peptide concentrations.

3.8 Water-Membrane Partition Experiments

For water-membrane partition experiments the Trp fluorescence signal of each peptide was followed titrating 1 μM peptide solutions with increasing amounts of 100 nm ePC/ePG or ePC/cholesterol liposomes. Spectra were collected between 300 and 450 nm using $\lambda_{exc.} = 280$ nm and a 295 nm cutoff filter in a 1 × 1 cm quartz cell. The average wavelength $\langle \lambda \rangle$ was calculated after background subtraction between 320 and 420 nm [2]. Anisotropy was measured for each peptide titrating a 1 μM peptide solution in the presence of increasing amounts of liposomes, using a $\lambda_{exc.} = 280$ nm and collecting the signal at 350 nm. No spectral corrections were needed in addition to blank subtraction in the lipid concentration range investigated, as checked by control experiments [2].

3.9 Molecular Dynamics Simulations of P5 and P5Del

MD simulations of peptides in water were performed in the laboratory where this thesis work was carried out, and are reported in this thesis for comparison with the experimental results. Trajectories started with the peptides in a helical conformation. The MD conditions were set as previously reported [8, 9], except for the pressure coupling that was applied isotropically. Five simulations, each 120 ns long, were performed for each peptide. Clustering was performed by considering the last 50 ns of each simulation (sampled with a 250 ps time step) by means of the *g_cluster* application available in the GROMACS software package [7], with default conditions. Molecular Lipophilicity Potential (MLP) [24] was calculated by using the Vega program [17] MD simulations of P5 in the presence of lipids was performed according to the "minimum bias" method [8, 9]. This simulation was 120 ns long, and a perfect bilayer formed in about 85 ns, after which the peptide position in the membrane remained stable. Molecular graphics were obtained using the UCSF Chimera package from the Resource for Biocomputing, Visualization, and Informatics at the University of California, San Francisco, or with the program VMD [23].

3.10 Molecular Dynamics Simulations of GAIV

The systems were simulated for 100 ns, adopting the conditions previously described [3, 15], except that in the present study semi-isotropic pressure coupling was used. Molecular graphics were obtained with the program VMD [9]. SLD profiles were calculated from the last 10 ns of the MD simulations, as follows: the density profile along the axis normal to the bilayer was calculated for each element of the system by using the g_density utility of the GROMACS package. These density values were multiplied by the coherent scattering length of the corresponding element [19] and then summed together to obtain the overall SLD. When multiple simulations were performed at the same peptide/lipid ratio, the SLD profiles were averaged over all available trajectories. These calculated SLD profiles cannot be compared directly to the experimental ones, since the latter have silicon on one side of the membrane, while in the simulations the bilayer faced a water phase on both sides. To overcome this issue, the experimental profiles were made symmetrical by reflecting the water-facing side with respect to the center of the hydrophobic region.

Order parameters for the for the C–C bonds of the palmitic chains were calculated with respect to the bilayer normal, according to standard definitions [3, 15], on the last 10 ns of the trajectories.

3.11 Neutron Reflectometry Experiments

Silicon single crystals ($5 \times 5 \times 1$ cm^3) cut along the (111) plane were cleaned by rinsing, under sonication, in chloroform, acetone, and ethanol. A 30 min UV/ozone treatment was used to make the polished surface more hydrophilic. Sample holders were laminar flow cells that allowed the injection of lipids and peptides and the exchange of the solvent to apply the well-known contrast variation method [5]. To this aim, high purity D_2O (ILL) and Milli-Q water were used with the following ratios 1:0, 0:1 or 38:62 (for silicon-matched water or SMW) in order to obtain media with scattering length density (SLD or ρ) 6.35×10^{-6} Å^{-2}, -0.56×10^{-6} Å^{-2} and 2.07×10^{-6} Å^{-2}, respectively. Neutron reflectivity experiments were performed at the silicon/water interface on the D17 reflectometer at ILL [6].

Two silicon monocrystalline solid supports were used and characterized before the injection of lipid vesicles. Solid supported POPC and (d_{31}) POPC lipid bilayers were obtained by the vesicle fusion method, which relies on spontaneous adhesion of the lipids to the surface following a vesicles spreading process [11]. Small unilamellar vesicles were formed sonicating a vesicle suspension (prepared as described in Sect. 3.3) by immersion of a titanium tip for three periods 10 min long at a power of 70 W, with suitable pauses between them to avoid excessive sample heating. Five mL of the vesicle suspension were injected in the sample flow cell. Excess of lipids not bound to the surface was removed by flushing water in the flow cell. To enhance the signal arising from the peptide molecules, addition of partially deuterated peptide was used in combination with hydrogenated POPC, while hydrogenous peptides were added to the partially deuterated lipid bilayer.

3.12 NR Data Analysis

The measured specular reflectivity data are connected to the arrangement of the material within the sample along the normal to the deposition interface. The relation can be summarized by the master formula [18]

$$R(Q_z) = \frac{16\pi^2}{Q_z^4} \left| \int \frac{d\rho(z)}{dz} e^{iQ_z z} dz \right|^2$$

where the compositional information is described by the SLD along the z direction, $\rho(z)$.

A limitation in the analysis of Neutron Reflectometry (NR) data is the impossibility of a direct Fourier transform of the reflectivity $R(Q_z)$ that would result in the SLD profile, as evident from the equation above. This limitation, known as phase-loss problem, can be overcome using the contrast variation method, i.e. modeling data originated from the same sample with different isotopic substitutions. The more are the contrasts available, the higher is the accuracy in the data modeling. For example, in the present experiment, all of the data available for a given concentration were modeled simultaneously to exploit this peculiarity of NR.

The data fitting procedure was based on the Parratt's recursion relation [16], originally derived for reflection of X-rays but nowadays widely used also for neutron scattering data. The simultaneous fit of the data was performed using an homemade software written in Matlab. The minimization routine was Fminuit, a χ^2 fitting program for Matlab based on the MINUIT minimization engine of the CERN program library [10]. The system was modeled as a stack of four layers, each of them described by a triplet of parameters [thickness t, $\rho(f_k)$ and interfacial roughness σ]. The parameter f_k is the volume fraction of the k component within the selected layer (see the equation reported below).

The first layer was used to describe the SiO_2 layer together with its hydration water. The sets of parameters, for both blocks, were obtained during the preliminary characterization of the bare silicon supports and kept fixed during the subsequent analysis. The remaining layers were respectively used to model: (i) the inner hydrophilic headgroup region (ii) the hydrophobic tail region and (iii) the outer headgroup layer. All of the information available about the sample components was introduced in the model as constraints.

If the composition of a layer is not uniform, i.e. in presence of more than one component, the SLD representing the layer can be expressed as a linear combination of the SLDs of the components, where the weighting factors are the component volume fractions. In general, for the ith layer the SLD was expressed as

$$\rho_i = f^i \rho_i^{dry} + f_w^i \rho_w + f_p^i \rho_p^{dry}$$

where f^i, f_w^i, f_p^i are respectively the volume fractions of the lipid, water, and peptide components in the ith layer, and their sum is normalized to unity.

References

1. Angelova MI, Dimitrov DS (1986) Liposome electroformation. Faraday Discuss Chem Soc 81:303–311
2. Bocchinfuso G, Bobone S, Palleschi A, Stella L (2011) Fluorescence spectroscopy and molecular dynamics simulations in studies on the mechanism of membrane destabilization by antimicrobial peptides. Cell Mol Life Sci 68:2281–2301
3. Bocchinfuso G, Palleschi A, Orioni B, Grande G, Formaggio F, Toniolo C, Park Y, Hahm KS, Stella L (2009) Different mechanisms of actions of antimicrobial peptides: insights from fluorescence spectroscopy experiments and molecular dynamics simulations. J Pep Sci 9:550–558
4. Chattopadhyay A, London E (1987) Parallax method for direct measurement of membrane penetration depth utilizing fluorescence quenching by spin-labeled phospholipids. Biochemistry 26:39–45
5. Crowley T, Lee E, Simister E, Thomas R (1991) The use of contrast variation in the specular reflection of neutrons from interfaces. Phys B 173:143–156
6. Cubitt R, Fragneto G (2001) Neutron reflection: principles and examples of applications. In: Pike ER, Sabatier P (eds) Scattering in microscopic physics and chemical physics, chap 2.8.3, pp 1–11
7. Daillant J (2005) Structure and fluctuations of a single floating lipid bilayer. Proc Natl Acad Sci USA 102:11639–11644
8. Heuber C, Formaggio F, Baldini C, Toniolo C, Müller K (2007) Multinuclear solid-state-NMR and FT-IR-absorption investigations on lipid/trichogin bilayers. Chem Biodivers 4:1200–1218
9. Humphrey W, Dalke A, Shulten K (1996) VMD: visual molecular dynamics. J Mol Graph 14:33–38
10. James F (1994) MINUIT minimization package, reference manual. CERN Program Library, Geneva
11. Kalb E, Frey S, Tamm LK (1992) Formation of supported planar bilayers by fusion of vesicles to supported phospholipid monolayers. Biochim Biophys Acta 1103:307–316
12. Kuipers BJH, Gruppen H (2007) Prediction of molar extinction coefficients of proteins and peptides using UV absorption of the constituent amino acids at 214 nm to enable quantitative reverse phase high-performance liquid chromatography-mass spectrometry analysis. J Agric Food Chem 55:5445–5451
13. Mazzuca C, Orioni B, Coletta M, Formaggio F, Toniolo C, Maulucci G, De Spirito M, Pispisa B, Venanzi M, Stella L (2010) Fluctuations and the rate-limiting step of peptide-induced membrane leakage. Biophys J 99:1791–1800
14. Merrifield B (1986) Solid phase synthesis. Science 232:341–347

15. Orioni B, Bocchinfuso G, Kim JY, Palleschi A, Grande G, Bobone S, Park Y, Kim JI, Hahm KS, Stella L (2009) Membrane perturbation by the antimicrobial peptide PMAP-23: a fluorescence and molecular dynamics study. Biochim Biophys Acta 1788:1523–1533
16. Parratt LG (1954) Surface studies of solids by total reflection of X-rays. Phys Rev 95:359–369
17. Pedretti A, Villa L, Vistoli G (2004) VEGA: an open platform to develop chemo-bioinformatics applications, using plug-in architectures and script programming. J Comp Aided Mol Design 18:167–173
18. Penfold J, Thomas RK (1990) The application of the specular reflection of neutrons to the study of surfaces and interfaces. J Phys: Condens Matter 2:1369–1412
19. Sears V (1992) Neutron scattering lengths and cross sections. Neutron News 3:26–37
20. Stella L, Mazzuca C, Venanzi M, Palleschi A, Didonè M, Formaggio F, Toniolo C, Pispisa B (2004) Aggregation and water-membrane partition as major determinants of the activity of the antibiotic peptide trichogin GA IV. Biophys J 86:936–945
21. Stella L, Pallottini V, Moreno S, Leoni S, De Maria F, Turella P, Federici G, Fabrini R, Dawood KF, Lo Bello M, Pedersen JZ, Ricci G (2007) Electrostatic association of glutathione transferase to the nuclear membrane. Evidence of an enzyme defense barrier at the nuclear envelope. J Biol Chem 282:6372–6379
22. Stewart JCM (1980) Colorimetric determination of phospholipids with ammonium ferrothiocyanate. Anal Biochem 104:10–14
23. Tielemann DP, Berendsen HJC, Sansom MSP (1999) An alamethicin channel in a lipid bilayer: molecular dynamics simulations. Biophys J 76:1757–1769
24. Woolf TB, Roux B (1996) Structure, energetics, and dynamics of lipid-protein interactions: a molecular dynamics study of the gramicidin a channel in a DMPC bilayer. Proteins 24:92–114

Chapter 4
Results and Discussion

4.1 Membrane Thickness and the Mechanism of Action of Trichogin GAIV

4.1.1 Trichogin GAIV

In this chapter the mechanism of action of the peptide trichogin GAIV (GAIV henceforth) will be investigated.

GAIV belongs to the family of peptaibols (or peptaibiotics), i.e. AMPs produced by the *Trichoderma* fungi. This class of pore-forming, bactericidal peptides, was actually discovered well before cationic AMPs [136]. They are produced non-ribosomally by fungi, are characterized by a C-terminal 1, 2-amino alcohol and a high content of non proteinogenic residues, most notably α-aminoisobutyric acid (Aib), and are usually acetylated or acylated at the N-terminus. Peptaibiotics are usually helical both in solution and when membrane-bound, and, unlike cationic AMPs, their content of charged residues is very low or even absent altogether. The best characterized member of this family (and the second to be identified, in 1967) is alamethicin (Alm) [76]. For this peptide it has been conclusively demonstrated through a combination of a large set of biophysical techniques that it forms pores through the "barrel-stave" mechanism [8, 19, 22, 36, 42, 74, 126]. This model was first proposed in 1974, but until now it has been convincingly demonstrated only for Alm.

Alm comprises 19 amino acids and its helix has a length corresponding almost exactly to the thickness of biological membranes (Fig. 4.1). However, it is also one of the longest members of the peptaibiotic family, which contains peptides going from 21 amino acids (SCH 643432) to just 4 (peptaibolin) [136]. Considering the closely related amino acid composition and physico-chemical properties of all members of the peptaibiotic family, it is conceivable that they could form pores in a similar way. However, an important drawback seems to preclude the formation of transmembrane pores to shorter peptides: how could such short helices form channels spanning through the whole bilayer?

S. Bobone, *Peptide and Protein Interaction with Membrane Systems*,
Springer Theses, DOI: 10.1007/978-3-319-06434-5_4,

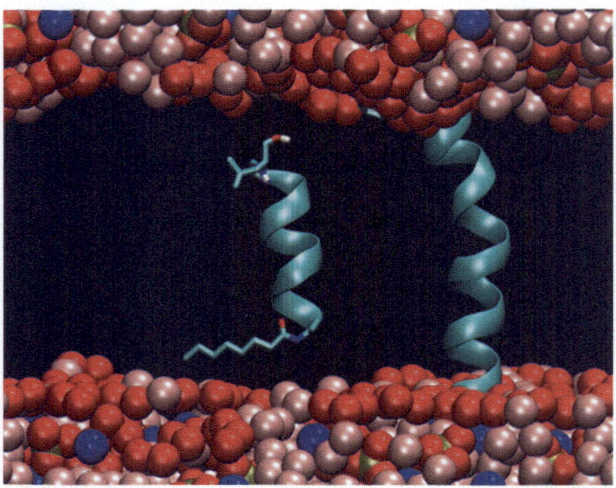

Fig. 4.1 Comparison of the length of the Alm and GAIV helices with the thickness of a POPC bilayer. Phosphorus, nitrogen and oxygen atoms of phospholipids are shown as *gold*, *blue* and *red spheres*, respectively, while water oxygen atoms are colored in *pink*. The acyl chains of the phospholipids are omitted for the sake of clarity. Peptides are represented as *ribbons*. The N-terminal acyl and C-terminal amino alcohol groups are shown in a "stick" representation [Reproduced from [15] with permission]

A well characterized, medium-length peptaibiotic is GAIV, whose sequence is

$$n\text{Oct - Aib}^1\text{ - Gly}^2\text{ - Leu}^3\text{ - Aib}^4\text{ - Gly}^5\text{ - Gly}^6\text{ - Leu}^7\text{ - Aib}^8\text{ - Gly}^9\text{ - Ile}^{10}\text{ - Lol},$$

where *n*Oct is *n*-octanoyl, and Lol is leucinol. This 10-mer peptide was isolated from *Trichoderma longibrachiatum* in 1992 [6] and since then it was studied both in solution and in model membranes by a number of physico-chemical techniques [108] including NMR [6, 54], X-ray crystallography [137], EPR [88–90, 114, 130, 131] fluorescence [45, 86, 126, 140–142] electrochemistry [10, 123], and molecular dynamics (MD) simulations [17]. Its 3D-structure is helical, with a flexible hinge in the central part, formed by two consecutive Gly residues [6, 137, 138, 141]. In this conformation, its length is only about half the normal thickness of a biological membrane (Fig. 4.1).

For this reason, different models were proposed to explain its pore-forming activity, including the SMH mechanism [38, 54, 90] and a carrier function [88, 89]. However, several evidences would favor a barrel-stave structure for its pores, just like Alm. In this connection, by combining several different fluorescence experiments it has been shown that, above a threshold concentration, GAIV inserts deeply into the hydrophobic core of the membrane and forms aggregates. These inserted, aggregated species are responsible for membrane leakage [45, 85, 126] Later on, these findings were confirmed by EPR measurements [114, 129, 130]. Other studies established that GAIV-induced membrane permeability is ion-selective and depends on the sign and magnitude of the transmembrane potential, exactly like that of Alm channels [10, 68, 122].

One hypothesis that was put forward to solve the apparent contradiction between these data and the short length of the GAIV helix is formation of head-to-head dimers able to span the entire bilayer [130, 137]. However, this hypothesis seems not to be supported by the voltage dependence of GAIV-induced membrane permeability, which would require a co-linear alignment of the helices forming the channel, to generate an overall dipole that could sense the transbilayer potential [10, 68] In a previous study [17] GAIV location inside a lipid bilayer was investigated by MD simulations using a self-assembling approach [20, 39] that has been defined as "minimum bias" [18, 99]. In this method, the simulation starts from a random mixture of lipids and water, containing one peptide molecule. During the simulation, a lipid bilayer self-assembles spontaneously, but in the initially highly fluid environment the peptide is able to attain its most favorable position in the membrane. This study [17] indicated a possible solution to the problem of the mismatch between membrane thickness and peptide length. In two out of three such simulations with GAIV, the peptide positioned close to the membrane surface, parallel to it, without significantly perturbing the membrane. However, in a third simulation it inserted into the bilayer in a transmembrane orientation. This simulation showed a bilayer that, near the peptide, was significantly thinned, so that the short GAIV helix was able to span completely the membrane. Nevertheless, this preliminary result could not rule out the possibility that the observed effect was an artifact of the way the bilayer formed, since in this simulation GAIV was initially inserted in a local defect of the membrane, which healed only after an extensive simulation. The short length of the equilibrated segment of that simulation (10 ns) left open the possibility that the observed bilayer structure was just a transient, metastable state, which would eventually relax. More importantly, this purely computational but intriguing indication needed an experimental verification.

In this chapter, combined experimental and simulative data will be reported, supporting the possibility for GAIV to form barrel-stave channels. Neutron reflectivity studies were used to experimentally verify the effects of GAIV on bilayer thickness. Moreover, vesicle leakage experiments were exploited to determine the influence of membrane thickness on GAIV activity. In addition, previous MD simulation studies were significantly extended, to confirm the stability of the transmembrane orientation, even when starting from a preformed bilayer, and to verify the cumulative effects of multiple membrane-inserted peptide chains. Overall, these data indicate that GAIV might be able to form barrel-stave channels by causing a significant thinning of the bilayer to a thickness comparable with the length of its helix (Fig. 4.1).

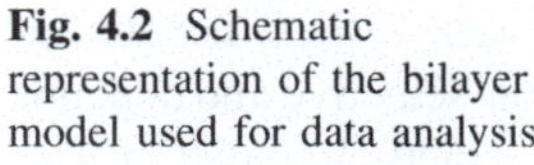

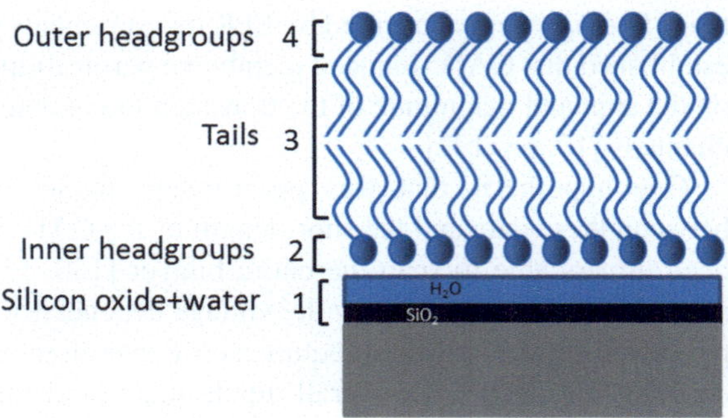

Fig. 4.2 Schematic representation of the bilayer model used for data analysis

4.1.2 Neutron Reflectivity Experiments

Specular neutron reflectivity measurements were carried out on planar POPC bilayers supported by a silicon crystal and submerged in a water phase, to which increasing GAIV concentrations were added. To increase the reflectivity contrast, in a first set of experiments, fully hydrogenated peptide was added to a chain-deuterated POPC bilayer (Fig. 4.3), while in a second set, partially deuterated GAIV (Fig. 4.4) was added to a normal POPC bilayer (Fig. 4.3) (See Table 4.1). Four different concentrations of peptide were used, namely C_0 (bare lipid bilayer), C_1 (4.5 μM), C_2 (15 μM), and C_3 (30 μM). All of the peptide-lipids-solvent combinations exploited are reported in Table 4.1. The data analysis has been carried out dividing the membrane into three sublayers, i.e. the outer headgroups region, the tails region, and the inner headgroup region. A further region is represented by the thin SiO_2 layer that covers the silicon support, with its hydration water (Fig. 4.2).

Each layer is characterized by defined values of SLD, thickness (d), and roughness (σ). The SLD of each layer will provide information about the species present in the considered region.

The cited parameters are used to calculate a model for reflectivity profile. The theoretical profile obtained by this model is then compared to the experimental one, and the quality of the fit is assessed calculating the χ^2 parameter in the least-squares method (as described in Sects. 3.11 and 3.12). The SLD and d for each layer can be varied until the optimum fit is found. All the reflectivity profiles obtained for the same peptide concentration under different contrast conditions were fitted simultaneously.

4.1.2.1 Bare Substrates Characterization

Two different Si substrates were used for the POPC and (d_{31})POPC bilayers. Before membrane deposition, these substrates were characterized using three different water compositions (Table 4.1). For that used in combination with hydrogenated POPC, the dioxide layer was characterized by a thickness $t_{ox} = 15.8 \pm 0.1$ Å,

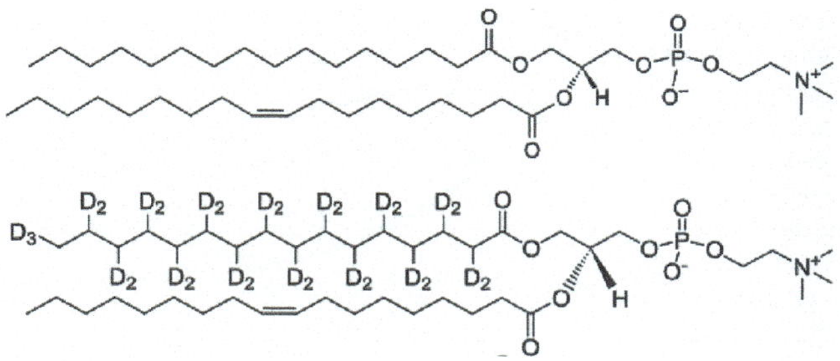

Fig. 4.3 Structures of hydrogenated (*top*) and deuterated (*bottom*) POPC

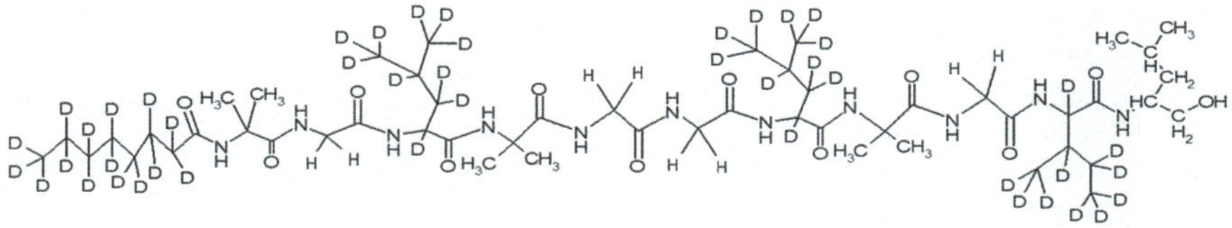

Fig. 4.4 Structure of the deuterated GAIV

Table 4.1 Datasets collected in the neutron reflectivity experiments

Lipids	Peptide	Concentration	D_2O	SMW	H_2O
POPC	--	C_0	X	X	X
(d_{31})POPC	--	C_0	X	X	X
POPC	deuterated	C_1	X		X
POPC	deuterated	C_2	X	X	
POPC	deuterated	C_3	X	X	
(d_{31})POPC	hydrogenated	C_1	X		X
(d_{31})POPC	hydrogenated	C_2	X	X	
(d_{31})POPC	hydrogenated	C_3	X		X

a volume fraction of hydrating water $f_{ox} = 0.15 \pm 0.05$ and a roughness $\sigma = 2.2 \pm 0.2$ Å. The block used for supporting deuterated lipids was characterized by the parameters $t_{ox} = 15.5 \pm 0.1$ Å, $f_{ox} = 0.12 \pm 0.05$, and $\sigma = 2.4 \pm 0.3$ Å. These values were kept fixed during the modeling of the sample data.

4.1.2.2 Bare POPC Bilayer Characterization

The simultaneous fits on all of the available datasets for a pure lipid bilayer are shown in Fig. 4.5. The parameters obtained are in complete agreement with those already reported in the literature [69]. In both cases, the parameters of the pure lipid

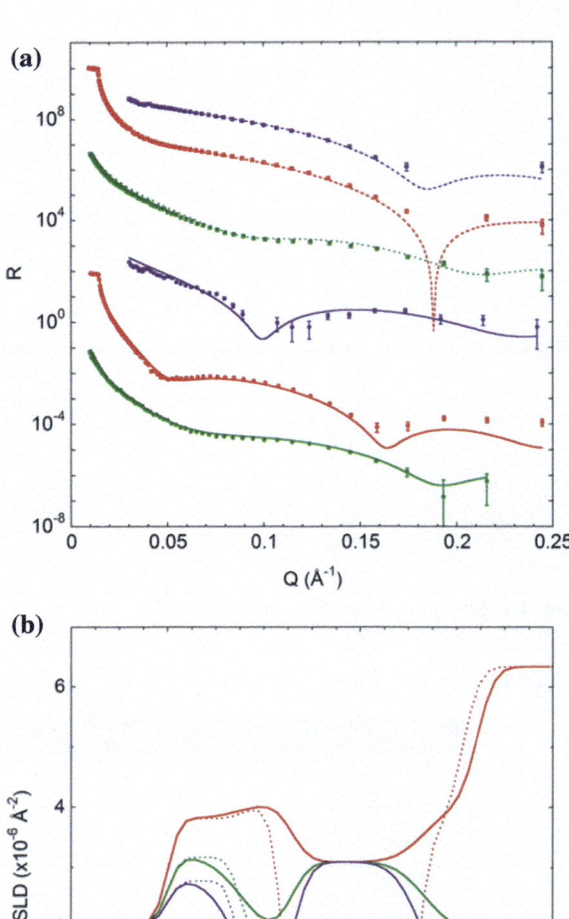

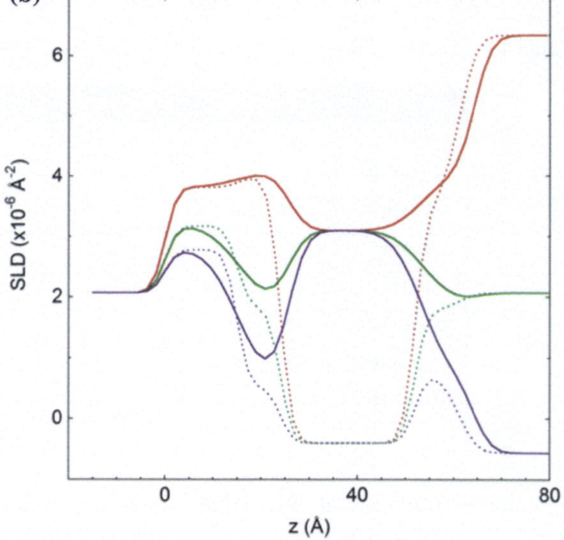

Fig. 4.5 *Panel A*: neutron reflectivity profiles for POPC and (d_{31})POPC bilayers in different media. From *top* to *bottom* POPC in H2O (*violet*), D2O (*red*), SMW (*green*) and (d_{31})POPC in H2O (*violet*), D2O (*red*) and SMW (*green*). Global fits to these data are reported as *solid* or *dotted lines*, for (d_{31})POPC and POPC data, respectively. The curves have been rescaled to improve the visibility. *Panel B*: SLD profiles for POPC and (d_{31}) POPC bilayers in different media. The color code is the same as in panel A [Reproduced from [15] with permission]

bilayer were indicative of a symmetric bilayer, with a total thickness of 48 ± 3 Å. The inner and outer headgroup regions were 10 Å thick (t_h = 10 ± 1 Å), with a volume fraction of water f_h = 0.5 ± 0.1. The hydrophobic core of the bilayer was 28 Å thick ($2xt_t$ = 28 ± 1 Å) with no water penetration, which suggests an almost perfect coverage of the substrate surface [28]. The roughness of all interfaces was similar and close to 2 Å. From these values the SLD profiles along the normal of the bilayer were evaluated (Fig. 4.5, bottom panel).

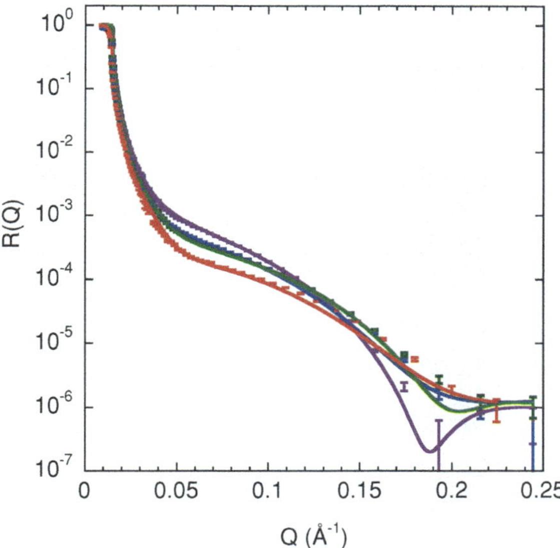

Fig. 4.6 Reflectivity profiles of POPC in D_2O. *Violet* no peptide; *Blue* C_1; *Green* C_2, *Red* C_3 [Reproduced from [15] with permission]

4.1.2.3 Peptide Effects

In Fig. 4.6 the curves of the POPC bilayer in D_2O at the different concentrations of deuterated peptide investigated are compared. Two main features are visible. From the pure lipid bilayer to the C_3 sample a decrease in reflectivity in the mid-Q region (0.05–0.15 Å^{-1}) is observed that could be interpreted as a change of contrast between the sample and the solvent. Indeed, the addition of deuterated peptide molecules into a hydrogenated bilayer would lead to a total SLD closer to that of D_2O. It has to be reminded that the contrast is the difference between the SLD of the sample and that of the medium. The lower is this value, the weaker is the reflectivity (or scattering) signal.

The second feature clearly visible is a shift of the main minimum of the profiles towards higher Q_z values. For the pure POPC bilayer it was located around $Q_z = 0.20$ Å^{-1} and it shifted to higher values after peptide insertion. This is a clear indication of thinning of the overall thickness of the deposition.

From the simultaneous fits according to the Parrat's recursive formula, the sample at the four different peptide concentrations was characterized in a clear and unambiguous way. The resulting parameters are listed in Table 4.2. From these parameters we were able to determine the main structural changes induced by the peptide inclusion into the lipid bilayer. First of all, the peptide is present only in the hydrophobic layer, as described by the volume fractions f_{ph} and f_{pt}, representing the volume fraction of peptide molecules occurring in the headgroup and tail regions, respectively. f_{ph} is zero for all of the investigated samples. Instead, the quantity of peptide inserting into the hydrophobic layer is clearly concentration dependent. From the peptide volume fraction f_{pt}, and from the molecular volumes

Table 4.2 Parameters obtained from data analysis of the lipid-peptide system at the four concentrations. t_h = thickness of headgroups region, t_t = thickness of tail region, f_h = volume fraction of water in the headgroup region, f_t = volume fraction of water in the tails region, σ_h = roughness of headgroups region, σ_t = roughness of tails region, f_{ph} = fraction of peptide in the headgroups region, f_{pt} = fraction N_{pt}/N_l = peptide to lipid ratio

	C_0	C_1	C_2	C_3
t_h (Å)	10 ± 1	9 ± 1	8 ± 1	9 ± 1
2 x t_t (Å)	28 ± 1	26 ± 1	23 ± 1	21 ± 1
f_h	0.5 ± 0.1	0.4 ± 0.1	0.4 ± 0.1	0.4 ± 0.1
f_t	0.00 ± 0.05	0.00 ± 0.05	0.05 ± 0.05	0.05 ± 0.05
σ_h *inner* (Å)	2 ± 1	2 ± 1	2 ± 1	2 ± 1
σ_h *outer* (Å)	2 ± 1	2 ± 1	2 ± 1	2 ± 1
σ_t (Å)	2 ± 1	2 ± 1	2 ± 1	2 ± 1
f_{ph}	n.a.	0	0	0
f_{pt}	n.a.	0.09 ± 0.02	0.11 ± 0.03	0.13 ± 0.05
N_{pt}/N_l	n.a.	0.07 ± 0.02	0.09 ± 0.02	0.10 ± 0.04

of GAIV and of the lipid tails the inserted peptide to lipid ratio at the different concentrations (N_{pt}/N_l) has been evaluated. Unfortunately, within the experimental accuracy, it was not possible to distinguish whether the peptide molecules are located in a specific part of the tail region or with a specific orientation with respect to the bilayer normal z.

The second important information arising from the modeling is that the inclusion of the peptide produces a thinning of the bilayer, especially because of a thickness decrease of the tail region. Actually, within the experimental accuracy, the headgroup thickness t_h is stable at all concentrations, while a decrease is observed in the t_t parameter (t_t is the thickness of the hydrophobic region of a single leaflet). The overall hydrophobic region is "compressed" by ~7 Å going from the pure lipid system to that with the highest amount of peptide included. This thinning affects the overall bilayer thickness, decreasing form 48 ± 3 to 39 ± 3 Å (Fig. 4.7). All of these structural changes are detectable also from a comparison of the SLD profiles, as shown in Fig. 4.8.

4.1.3 Molecular Dynamics Simulations

In order to understand at the molecular level the structural changes observed in the bilayer in the NR experiments, MD simulations of GAIV-membrane systems were performed in the research group where this thesis was carried out. In a previous computational study [17] it has been shown that when the peptide is associated to the membrane parallel to its surface it does not cause any significant bilayer perturbation. Therefore, in the present study the stability of a transmembrane peptide orientation and its effects on the bilayer has been investigated.

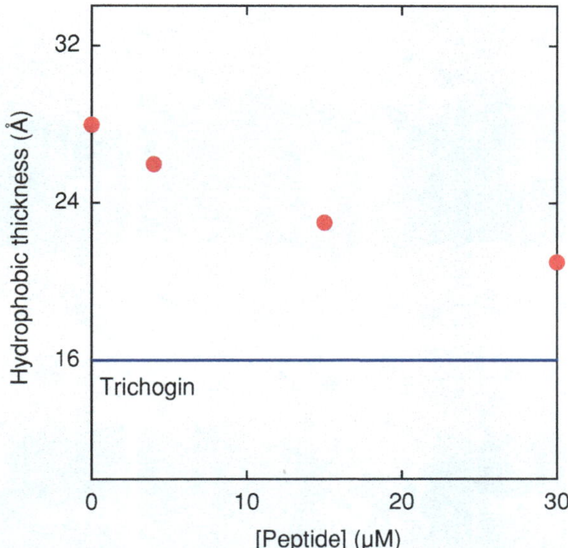

Fig. 4.7 Hydrophobic thickness versus peptide concentration. The *blue line* represents GAIV length [15]

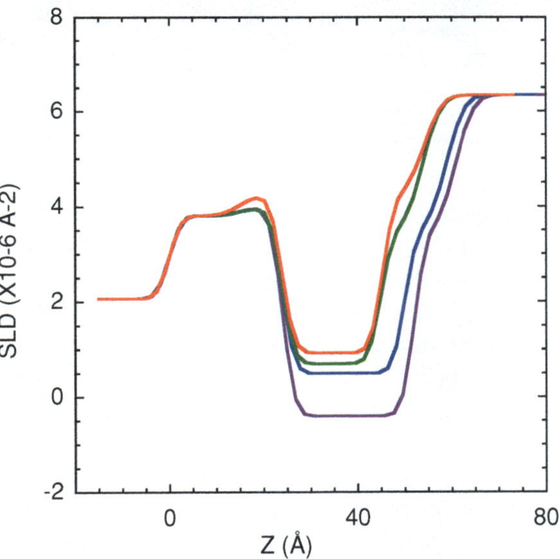

Fig. 4.8 SLD profiles of POPC in D_2O. The colors code is the same of Fig. 4.6 [Reproduced from [15] with permission]

The peptide was inserted in the bilayer according to a protocol developed specifically for this purpose [15, 16]. After GAIV insertion and an appropriate 10 ns equilibration period (during which the peptide was position-restrained), the simulations were continued for 100 ns more. A total of 7 simulations was performed: in one of them, only one peptide molecule was inserted in the bilayer (comprising 128 POPC lipid molecules), in three other simulations 4 peptide

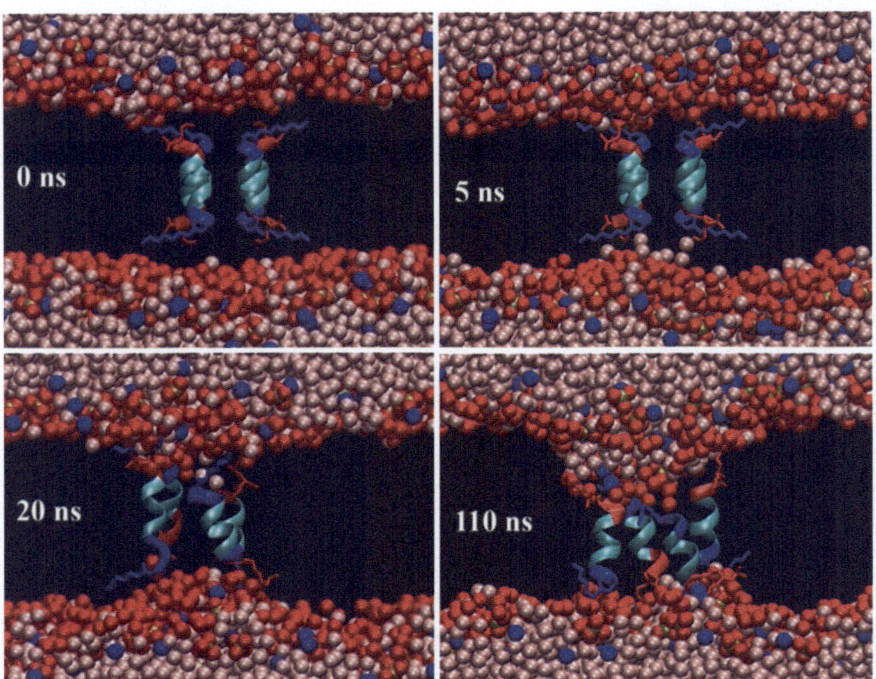

Fig. 4.9 Snapshots showing the evolution of the structures during the simulation with four GAIV molecules [Reproduced from [15] with permission]

molecules were inserted, and in further three simulations the peptides in the bilayer were 8. In most cases, the peptide molecules maintained a transmembrane orientation for the whole trajectory, and quickly (i.e. in about 10–30 ns) some lipid headgroups in the region above and below the peptide were drawn deeper in the membrane, due to the interaction with the free peptide NH and CO groups at the two termini of the helices. A few structures representative of the time-evolution of the simulations are shown in Fig. 4.9, and the structures at the end of the simulations are illustrated in Fig. 4.10.

To better define the interactions responsible for the observed bilayer deformation an analysis was carried out of the interactions of the peptide NH and CO groups not involved in intramolecular H-bonds, and of the OH group of the C-terminal amino alcohol, with different parts of the lipid molecules or with water. This analysis was limited to the trajectory segment following the formation of the bilayer thinning, i.e. from 30 ns onward (Fig. 4.11). As expected, the NH groups at the N-terminus interact mainly with the oxygen atoms of the phosphate and glycerol groups, while the C-terminal CO groups were almost invariably associated with the quaternary ammonium of the lipid choline group. The OH group of Lol interacts mostly with the glycerol moiety. All three peptide groups (NH, CO and OH) also interact with water molecules. These data indicate that the bilayer

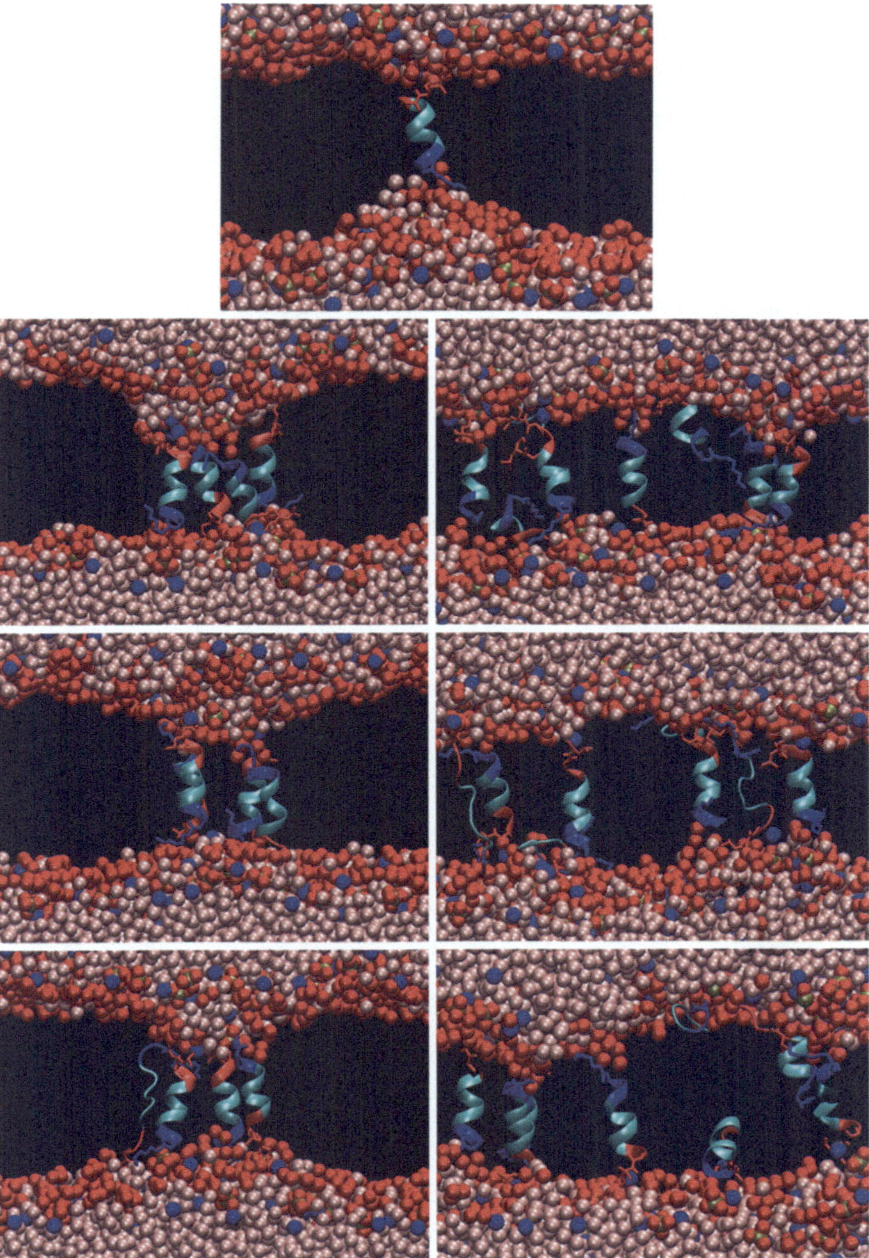

Fig. 4.10 Final structures obtained from simulation with, 1, 4 or 8 peptide molecules [Reproduced from [15] with permission]

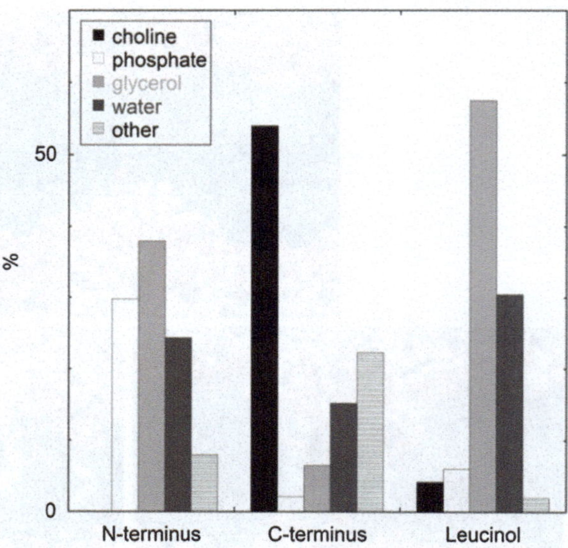

Fig. 4.11 Statistics of the interactions of selected peptide groups during the equilibrated segments of the MD trajectories with 4 GAIV molecules (30–110 ns). For each trajectory frame, the lipid or water atom closest to the different peptide groups (N-terminal NH, C-terminal CO and amino alcohol OH) was identified [Reproduced from [15] with permission]

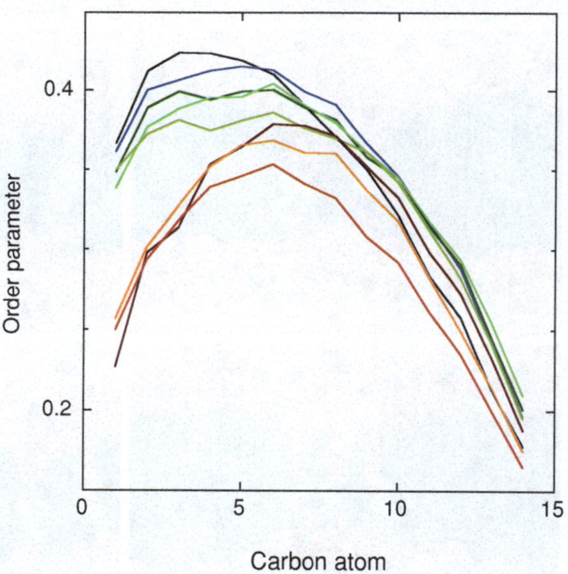

Fig. 4.12 Order parameters for the C–C bonds of the palmitic chains, calculated with respect to the bilayer normal, on the last 10 ns of the trajectories. *Black*: no peptide (*black*), *blue*:1, *green*: 4, *red*: 8 peptide molecules [Reproduced from [15] with permission]

deformation is driven by electrostatic and H-bonding interactions, which are enhanced in the low-dielectric constant environment of the hydrophobic membrane core. These interactions lead to thinning by drawing phospholipid headgroups deep into the membrane, thus inducing an increase in the number of gauche

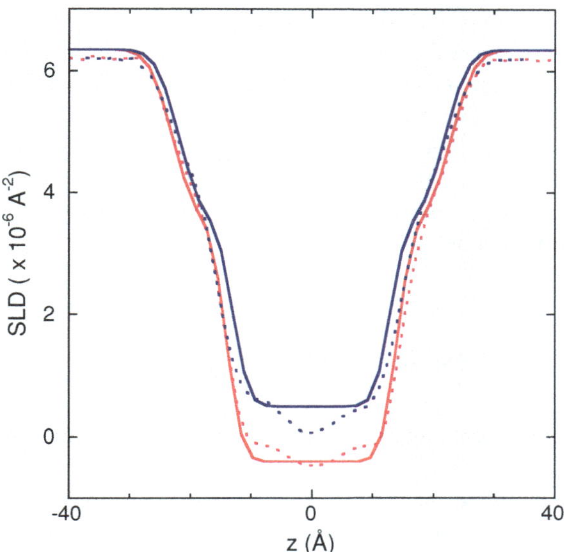

Fig. 4.13 Comparison between the neutron SLD profiles calculated from the MD trajectories of a peptide-free POPC bilayer (*pink dotted line*) and in the presence of 8 trichogin molecules (*violet dotted line*, peptide to lipid ratio 0.0625) with the symmetrical profiles obtained from the experimental neutron data of the peptide-free bilayer (*pink solid line*) and in the presence of deuterated GAIV at concentration C1 (*violet solid line*, estimated inserted peptide to lipid ratio 0.07 ± 0.02) [Reproduced from [15] with permission]

conformations in the lipid tails, as shown by a progressive decrease in their order parameters, as the number of peptides in the bilayer increases (Fig. 4.12).

To further confirm the stability of the transmembrane state, the simulation previously obtained by using the "minimum bias approach", whose equilibrated segment was initially just 10 ns long, was extended by further 40 ns. During this time, no significant modification in the peptide location, nor in the bilayer structure in its surroundings was observed (data not shown). The same system was simulated at 350 K for 65 ns without any evidence of destabilization of the transmembrane conformation.

Overall, these simulations strongly support the possibility of a transmembrane orientation for GAIV and a significant thinning effect on the bilayer, so that the short peptide helix could span it from one side to the other. However, in order to assess the reliability of the MD results, it was essential to evaluate their agreement with the experimental data.

4.1.3.1 Comparison with the NR Data

Neutron SLD profiles can be calculated from the MD trajectory and compared with the experimental data (Sect. 3.9). However, in order to perform a meaningful comparison, it was necessary to take into account the fact that in our experiments the lipid bilayer was supported on a silicon crystal, while it was free-standing in water in the simulations. It was previously demonstrated that GAIV very quickly partitions in both layers of the membrane [85] and this should lead to a symmetrical SLD profile for a trichogin-containing bilayer in water.

Therefore, the SLD profile of a free-standing membrane was obtained from the experimental profiles by reflecting the section of the profile corresponding to the water-facing side of the bilayer with respect to the center of the hydrophobic region. The comparison of the resulting profiles, reported in Fig. 4.13, shows a good qualitative agreement, regarding both the thickness of the bilayer and the increase in SLD due to peptide insertion in the membrane. This comparison confers a good confidence to the atomic-level picture provided by the MD simulations.

4.1.4 Vesicles Leakage Experiments

Both the simulations and the neutron reflectivity experiments indicate the possibility for GAIV to span the bilayer entirely, by causing a significant thinning of the membrane. However, this bilayer deformation requires a free energy cost. Therefore, if a transmembrane orientation is involved in the GAIV pore-formation process, the membrane-perturbing activity of this peptide should increase significantly in thinner membranes, which require a smaller deformation or no thinning at all. To verify this point, peptide-induced vesicle leakage experiments have performed with liposomes formed by lipids with different chain lengths and bilayer thicknesses [55] (Figs. 4.14 and 4.15). For comparison, the same experiments were carried out also with the much longer peptaibol Alm (Fig. 4.16). As shown in Figs. 4.15 and 4.16, the activity of GAIV increases dramatically with decreasing the bilayer thickness, while that of Alm is affected only marginally. In bilayers with a thickness comparable to that of biological membranes (i.e. with a number of carbon atoms of 16–18) the activity of GAIV is much lower than that of Alm. However, in the thinner membranes (whose hydrophobic core has a size comparable to that of the GAIV helix), the activity of the two peptides becomes comparable (Fig. 4.17). It is also worth mentioning that the curves of peptide-induced leakage as a function of GAIV concentration (Fig. 4.15) are steeply sigmoidal in membranes with high thickness, indicating a strong cooperativity of the pore formation process, but this cooperativity decreases drastically in thinner membranes.

4.1.5 Discussion

Both NR data and MD simulations show a peptide-induced thinning of the membrane. The effect of the insertion of a peptide/protein in a bilayer with an equilibrium hydrophobic thickness differing from that of the inclusion has been rationalized in terms of hydrophobic mismatch [63, 92]. The free energy cost of exposing to water hydrophobic groups of the protein or lipids, due to their different thickness, is higher than that of distorting the lipids from their equilibrium

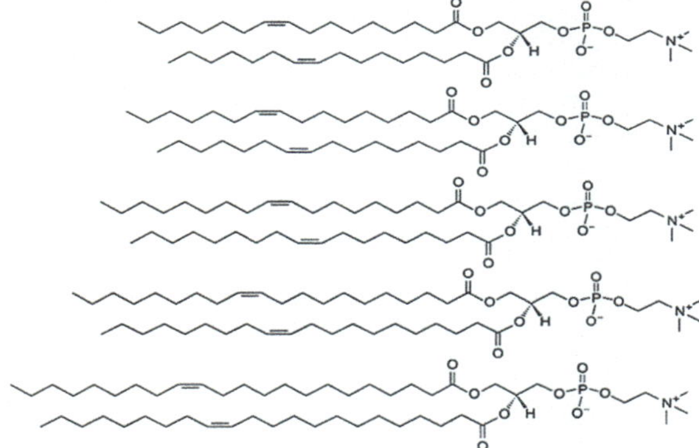

Fig. 4.14 Structures of the lipids used for the leakage experiments

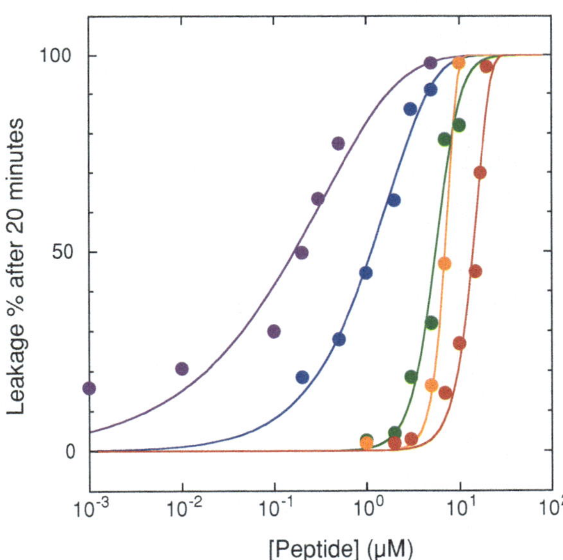

Fig. 4.15 Leakage curves of GAIV (A) in vesicles with a different hydrophobic thickness: *Red* di22:1 PC (1, 2-dierucoyl-sn-glycero-3-phosphocholine); *orange* di20:1 PC (1, 2-dieicosenoyl-sn-glycero-3-phosphocholine); *green* di18:1 PC (1, 2-dioleoyl-sn-glycero-3-phosphocholine); *blue* di16:1 PC (1, 2-dipalmitoleoyl-sn-glycero-3-phosphocholine); *violet* di14:1 PC (1, 2-dimyristoleoyl-sn-glycero-3-phosphocholine) [Reproduced [15] with permission]

conformation. As a consequence, the membrane thickness locally adapts to the size of the inclusion, although other effects are also possible [64]. The case of trichogin falls under the category termed *negative mismatch*, where the inclusion is shorter than the membrane thickness. A systematic study on model peptides demonstrated that hydrophobic mismatch is sufficient to drive membrane thickness adjustments comparable to those that would be necessary for barrel-stave pore formation by trichogin [67]. However, the MD results indicate that, in the case of trichogin, membrane thinning might be driven also by specific interactions between the peptide N- and C-termini and the phospholipid headgroups. In this respect, it is important to note that we have previously shown that in solvents of low polarity trichogin binds cations with a very high affinity [142]. This result might reflect

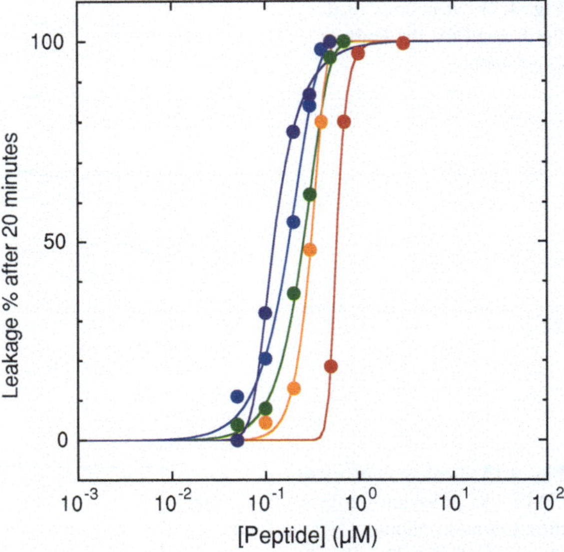

Fig. 4.16 Leakage curves of Alm (B) in vesicles with a different hydrophobic thickness: The color code is the same than Fig. 4.15 [Reproduced from [15] with permission]

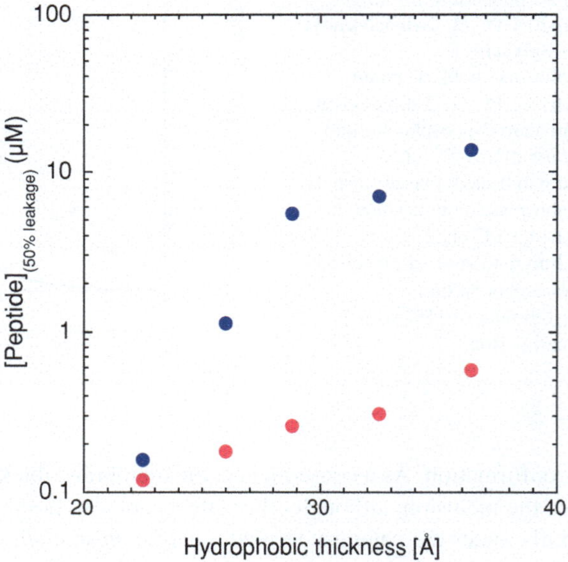

Fig. 4.17 Membrane-permeabilizing peptide activity as a function of the hydrophobic thickness of the bilayer. The peptide concentration is required to induce the leakage of 50 % of liposome contents is reported on y axis [15]

what is happening in the interaction between the peptide and lipid headgroups in the low dielectric environment of the bilayer core.

A completely different interpretation for peptide-induced membrane thinning has been provided by Huang [58] and used to describe the mechanism of action of both cationic AMPs and peptaibiotics. According to his model, binding of an amphipathic molecule to the bilayer surface at the water-lipid chain interface leads to an increase in the interfacial area, and thus to a decrease in the hydrocarbon thickness, due to the very low volume compressibility of the lipid chains. This deformation has an elastic energy penalty, and thus, at a threshold bound peptide concentration, a transmembrane orientation becomes favored. In this model, membrane thinning is due to the surface-bound peptide, rather than to the transmembrane-inserted molecules. Unfortunately, previous works showed that in both orientations the peptide is located essentially in the hydrophobic core of the membrane [17, 85, 126] and therefore the NR data do not allow us to discriminate between the Huang hypothesis and thinning due to transmembrane inserted peptides. However, several indications suggest that the Huang mechanism is unlikely for trichogin: insertion of this highly hydrophobic peptide in the phospholipid tail region, parallel to the membrane surface, can be easily accommodated, without causing a significant increase in the interfacial area nor a perturbation of membrane order, as also shown by previous simulations and experiments performed by our research group [17].

Whatever the driving force of the trichogin-induced membrane thinning might be, this effect makes possible for the peptide to span the bilayer from one side to the other, and also provides a rationalization to a previous observation regarding this peptaibol. Fluorescence experiments demonstrated that membrane-inserted trichogin exists essentially in an aggregate state [85, 126]. Clearly, the free energy cost of membrane deformation can be significantly reduced by peptide aggregation (just like in the hydrophobic effect aggregation of apolar molecules in water reduces the entropic cost of water structuring around them). Therefore, a monomeric transmembrane mismatched inclusion is significantly unstable [64]. This is probably the main driving force for the highly cooperative oligomerization in membrane-permeabilizing channels. This interpretation would also explain the observation that curves of membrane-perturbing activity as a function of peptide concentration are highly cooperative in thick membranes, while this cooperativity decreases significantly in thinner membranes (Fig. 4.15).

Alm is one of the longest peptaibiotics, but many "medium"- or "short"-length peptides of this class do exist [30, 136]. Therefore, the present findings might be relevant for a rather wide class of peptides. Longer peptaibols are much more active than shorter ones [46], but barrel-stave channel formation has been hypothesized for some of the latter, like the 16 residue long antiamoebin [37]. Solid-state NMR measurements have also shown that such relatively short peptides, like the 14 residue ampullosporin or the 15 residue zervamicin II, attain a predominantly transmembrane orientation when the membrane thickness is comparable to their length, while they are largely parallel to the surface in thicker bilayers [9, 115]. Therefore, it has been proposed that all peptaibiotics might act according to the barrel-stave mechanism, the lower activity of the shorter ones being due to the lowest fraction of peptide molecules in a transmembrane orientation [115].

Some other findings reported in the literature further support the possibility for a peptaibiotic to form barrel-stave channels in a membrane thicker than its length. For instance, Alm is able to form pores even in artificial membranes formed by diblock copolymers whose hydrophobic region is much thicker than the peptide length [143], thus paralleling the situation of trichogin in biological membranes. In addition, our findings regarding the membrane thickness dependence of peptide's activity nicely parallel the observation that in a series of short-chain trichogin analogs the activity significantly and progressively decreased with the shortening of the peptide chain [41].

No electrophysiology measurements of the conductance of single trichogin channels in planar membranes, which would provide a conclusive confirmation of a barrel-stave mechanism, have been reported to date. However, experiments on liposomes have shown that the pores formed by this peptide are ion selective and that their conductance is voltage-dependent, as it would be expected for Alm-like barrel-stave pores [68]. The voltage-gated nature of the trichogin channels was recently confirmed also by electrochemical measurements on Hg-supported tethered bilayer lipid membranes [10]. Overall, the present data indicate that formation of transmembrane barrel-stave channels might indeed be possible even for short peptaibiotics.

4.2 Selectivity

4.2.1 Role of Pro Residues in AMPs Sequences

In order to be able to design new peptide-based antibiotic drugs, it is essential to define the peptide characteristics that are responsible for AMPs selectivity.

The importance for selectivity of some AMPs molecular characteristics, such as charge and amphiphilicity, for their selectivity has been discussed in the Chap. 1. In the present chapter the role of another feature, common to many AMPs, will be discussed, i.e. the presence of a Pro or Gly residue in the amino acidic sequence. It is a conserved feature, present in several peptides [138, 157]. A statistical analysis quantitatively demonstrated the frequent presence of Gly residues close to the peptide center [138], but a similar statistical evaluation has not been reported for Pro. Nevertheless, several examples of Pro-containing helical AMPs or membrane-active toxins are described in the literature: Alm [127], BMAP-27 and BMAP-28 [147, 151], caerin [147], cecropin [156], fowlicidin [148], gaegurin [129], indolicin [151, 152], maculatin [26], melittin [31], pardaxin [146], pin2 [113], PMAP-23 [99], SMAP-29 [132], temporin A [23], tripticin [149], XT-7 [128], and an antimicrobial lysozyme fragment [60], among others. For many of these a Pro-induced kink in the helical structure has been demonstrated by NMR studies in solution or in membrane-mimicking media.

An analysis carried out using the AMPs database APD2 (http://aps.unmc.edu/AP/main.php), showed that in AMPs with less than 25 residues, and a single Pro residue, this amino acid is mainly found in proximity of the sequence center, with a preference for position closer to the N-terminus (Fig. 4.18).

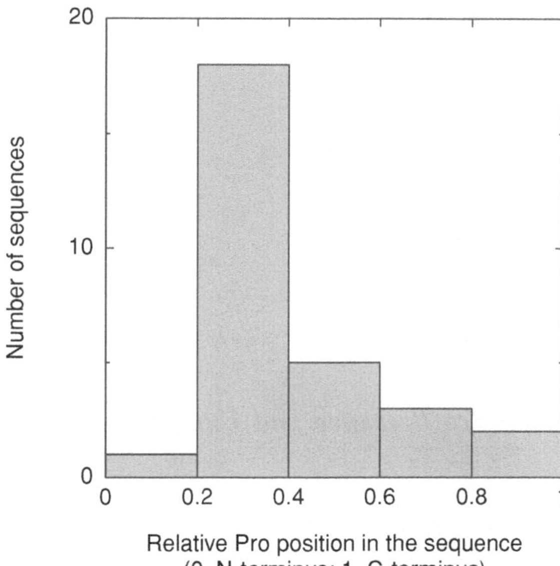

Fig. 4.18 Statistics of Pro position in helical AMPs, 11–25 residues long, containing a single Pro residue [Reproduced from [16] with permission]

The possible role of Pro in AMPs activity or selectivity is less easy to understand, with respect to the positive charge, or amphipaticity, which can be immediately related to the association to the lipid membranes. To study the effect of the presence and of the position of Pro in AMPs sequences, the P5 peptide was chosen.

4.2.2 The P5 Peptide

P5 is a model peptide, designed with the aim to obtain new synthetic peptides with a strong antibacterial activity. It was first derived from an hybrid parent peptide, CA-MA [104]. It is a Leu-Lys rich peptide, with a strong antibacterial activity and high selectivity, being nontoxic for erythrocytes up to the maximum investigated concentration, corresponding to 100 μM, i.e. 100-fold higher than the one required to kill bacteria [104]. The sequence of P5 is the following:

KWKKLLKKPLLKKLLKKL – NH_2

The peptide sequence clearly shows some of AMPs characteristics, like the positive net charge. If a helical structure is assumed for P5, we can notice how the amphipathic sequence has been designed to obtain two portions of a perfect amphipathic helix.

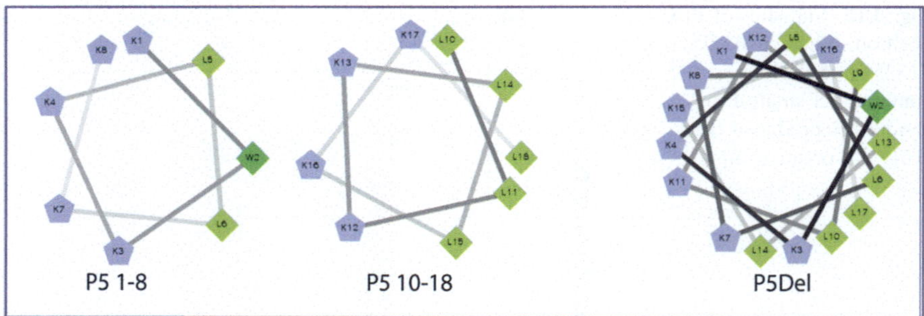

Fig. 4.19 Helical wheel projections of P5 and P5Del [16]

4.2.3 Pro Presence and Function in Proteins

The conformational effects induced by Pro in helical conformations are due to its particular structure: Pro is the only N-substituted residue, with a cyclic side chain. Due to this characteristics, Pro has a restricted torsional space, and its backbone amide, lacking a proton, cannot act as a donor in H-bond; thus, a Pro in i position will break the i to $i + 4$ H-bond usually formed in alpha-helices, and also the $i + 1$ to $i - 3$, for steric reasons [111]. Furthermore, it has been shown that Pro perturbs the conformational space of the preceding residue [59, 70, 118]. For all these reasons, it can dramatically affect the stability of helical structures, when located in the middle of the amino acidic sequence.

As a consequence, in soluble proteins Pro can often be found in proximity of the termini, but rarely in the helix interior [70, 111]. It has been estimated that only 5 % of total Pro residues are found inside helices, and that in this case they induce a kink [7, 80], conferring high flexibility to the protein structure.

Since in helical regions of soluble proteins Pro is extremely rare, and it is often located in hydrophilic regions, it would not be expected to be found in transmembrane helices. By contrast it surprisingly represents about 3–4 % of TM helical residues [77, 116], and, in addition, it is preferentially located close to the center of the helix [27, 117, 139]. Like in soluble proteins, also in the membrane environment Pro induces a kink [135] and indeed about 64 % of TM helices are kinked. Pro is usually 2 or 3 residues in the C-terminal direction with respect to the bend [73]. Pro residues are usually rather conserved in TM helices, hinting to a possible functional role [21, 117, 139]: it has been hypothesized that the Pro-induced kink could be needed to have free CO groups inside membrane channels, that could participate in ligand binding, or to gate channels with cis/trans isomerization or Pro-induced flexibility [21, 35, 78].

In the case of AMPs, no clear explanation has been proposed for the presence of the central kink. P5 is an excellent model system to study the role of Pro residues in AMPs.

Table 4.3 Amino acidic sequences of P5 and its analogues

P5 (Pro9)	**KWKKLLKKPLLKKLLKKL-NH_2**
P5Pro5	KWKK**P**LLKKLLKKLLKKL-NH_2
P5Pro7	KWKKLL**P**KKLLKKLLKKL-NH_2
P5Pro11	KWKKLLKKLL**P**KKLLKKL-NH_2
P5Pro13	KWKKLLKKLLKK**P**LLKKL-NH_2
P5Pro15	KWKKLLKKLLKKLL**P**KKL-NH_2
P5Del	KWKKLLKK-LLKKLLKKL-NH_2

4.2.4 P5 Analogues

In order to perform a systematic and complete study, a series of analogues were designed and synthesized starting from P5. In every analogue, Pro was moved towards the peptide termini, in positions which would keep the amphipathic character of the two helix portions, or removed altogether. In this case, a perfect amphipathic helix was obtained (Fig. 4.19).

The structures of P5 analogues are summarized in Table 4.3.

4.2.5 Antimicrobial and Hemolytic Activity

The antimicrobial activity of P5 and its analogues was tested against different Gram-positive and Gram-negative bacteria. The bactericidal activity is expressed as minimum inhibitory concentration, or MIC, defined as "the lowest concentration of an antimicrobial that will inhibit the visible growth of a microorganism after overnight incubation" [4].

All analogues featured a good antimicrobial activity, even though the Pro displacement caused a slight increase in MIC values, that are shown in Fig. 4.20.

The hemolytic activities of the analogues, instead, are dramatically affected by the modification of Pro position: the results show a strong increase in the peptides' toxicity as Pro is shifted from the sequence center or removed altogether. P5 is not hemolytic up to the higher tested concentration, 100 μM [104]. By contrast, analogues have an increasing toxicity, which becomes stronger as the Pro residue is shifted along the sequence. The most toxic analogue is P5Del, which has a hemolytic activity comparable to that of melittin (a potent bactericidal peptide

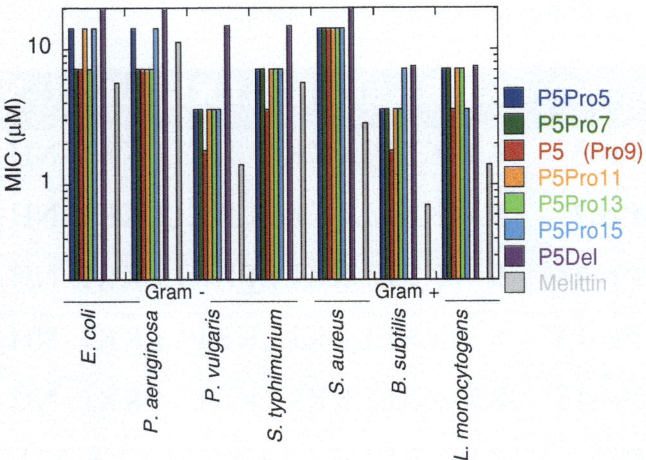

Fig. 4.20 Antibacterial activity of P5 analogues on different bacterial strains

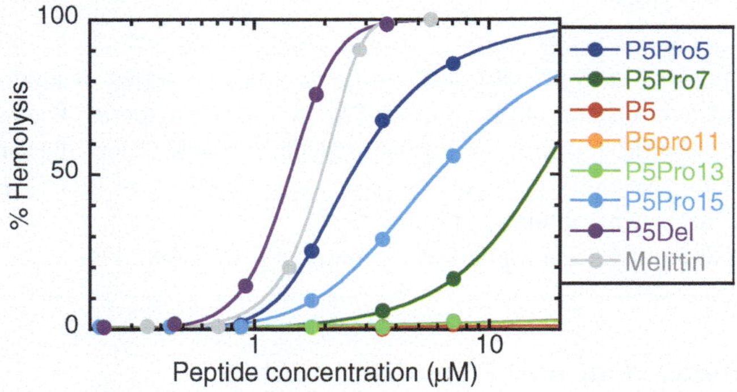

Fig. 4.21 Hemolytic activity of the analogues

derived from the bee venom, which is also dramatically hemolytic) as shown in Fig. 4.21.

The dramatic differences in peptides selectivity could depend by a different affinity for membranes of different composition, or by a different perturbing activity towards different membranes. Studies on model membranes were performed to clarify this point.

P5, the parent peptide, and P5Del, the Pro-lacking peptide. show the greatest differences in hemolytic activity. Therefore, these two AMPs have been chosen to perform experiments on model membranes.

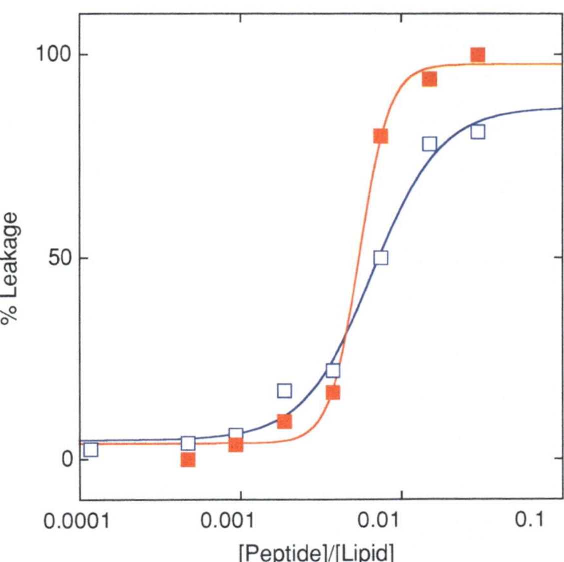

Fig. 4.22 CF leakage from ePC/ePG vesicles. *Red, filled symbols*: P5. *Blue, empty symbols*: P5Del [16]

4.2.6 Membrane-Perturbing Activity of P5 and P5Del

4.2.6.1 Peptide-Induced CF Leakage

To perform carboxyfluorescein leakage, to investigate the membrane-perturbing ability of P5 and P5Del [18], 0.03 μM peptide was added to various liposome solutions of different lipid concentrations.

In ePC/ePG (2:1 molar ratio) liposomes P5 and P5Del show a very similar activity (Fig. 4.22), causing a strong membrane perturbation and both reaching a value close to 100 % of leakage in a similar range of P/L values.

In ePC/cholesterol (1:1 molar ratio), on the other hand, P5Del resulted to be dramatically more active than P5 (Fig. 4.23).

To investigate the role of cholesterol, which has been suggested as a possible cause for the low activity of AMPs towards eukaryotic membranes, as discussed in Sect. 1.4 of Chap. 1 [84], the same experiments were performed also using vesicles composed only by ePC: also in this case the activity of P5 was significantly lower than that of P5Del (Fig. 4.24). This result proved that the differences in activities are due to the bilayer negative charge, given by ePG phospholipids, and not to the presence of cholesterol, which has only a slight influence on P5 behavior. The leakage data are in good agreement with the findings obtained by the hemolytic and antibacterial activity assays, in which the two peptides showed a similar bactericidal power, but only P5Del resulted to have hemolytic properties.

These experiments also allowed to point out that the distinct biological activities of the two analogues derive from their different ability to perturb membranes of different composition.

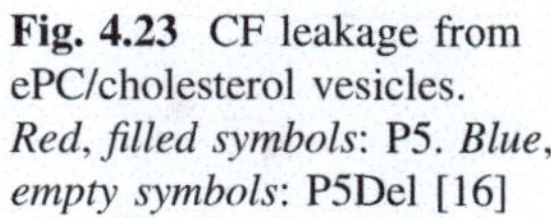

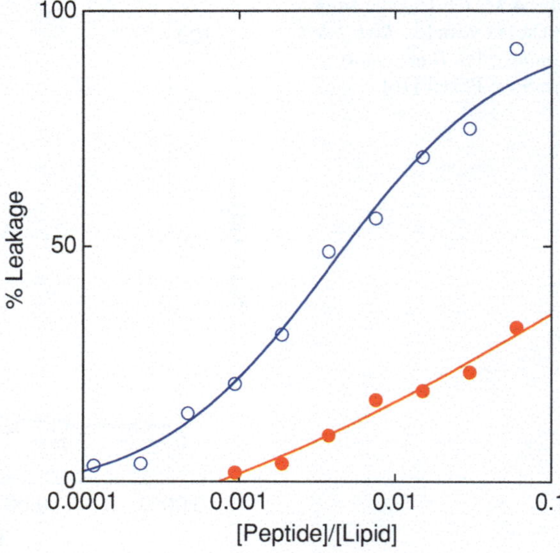

Fig. 4.23 CF leakage from ePC/cholesterol vesicles. *Red, filled symbols*: P5. *Blue, empty symbols*: P5Del [16]

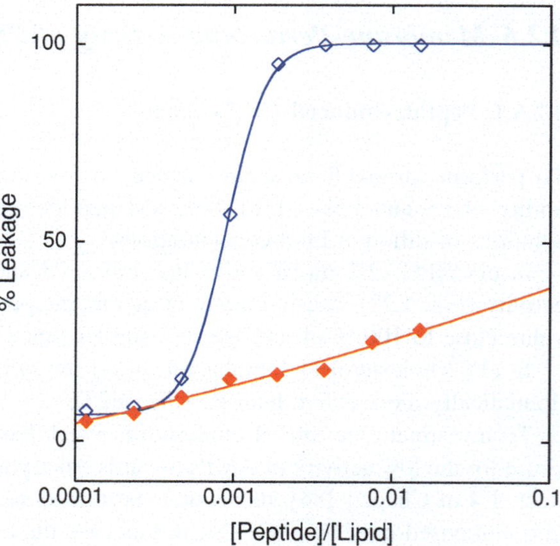

Fig. 4.24 CF leakage from ePC vesicles. *Red*, filled symbols: P5. *Blue*, empty symbols: P5Del [16]

The effect of cholesterol resulted to be negligible; for this reason, all the following experiments on peptide-induced membrane-perturbation were performed using PC/PG and PC/cholesterol liposomes only.

4.2.6.2 Vesicles Aggregation

Liposomes-induced aggregation was investigated measuring the light scattering increment at 400 nm. The experiments were performed titrating P5 or P5Del

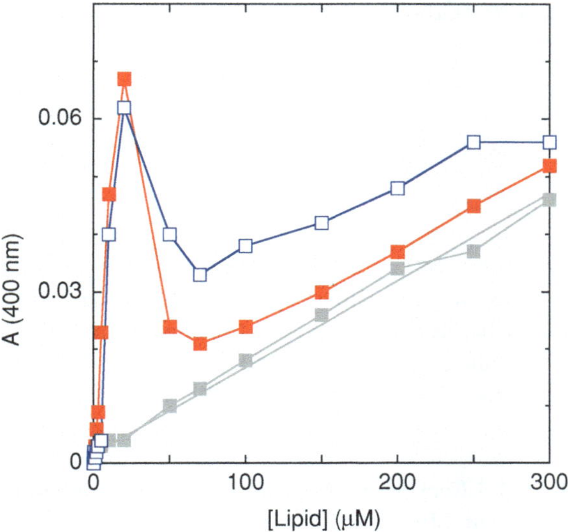

Fig. 4.25 Peptide-induced ePC/ePG vesicles aggregation. *Red*: P5, *blue*, P5Del, *grey*: no peptide [16]

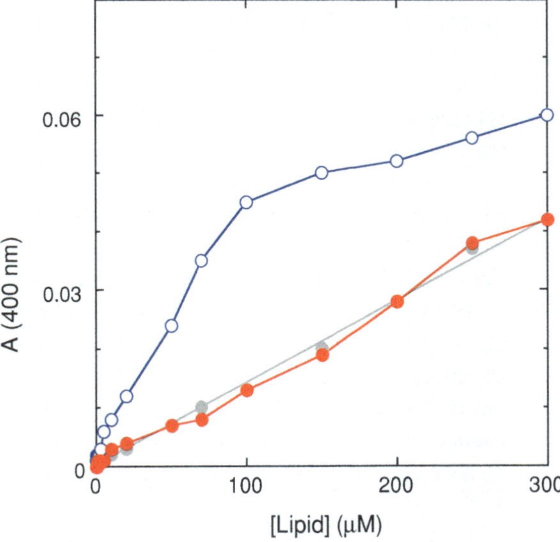

Fig. 4.26 Peptide-induced ePC/cholesterol vesicles aggregation. *Red*: P5, *blue*, P5Del, *grey*: no peptide [15]

(at 1 μM concentration) with increasing lipid amounts. If liposomes start to aggregate, an increase in the turbidity of the suspension occurs. P5 and P5Del are able to induce vesicle aggregation, as measured by the turbidity of liposome suspensions. For both P5 and P5Del, the turbidity of a peptide-containing suspension of charged vesicles is significantly higher than that of peptide-free liposomes, indicating peptide-induced aggregation (Fig. 4.25). Interestingly, this phenomenon is largely reversible after a given lipid to peptide ratio is reached, indicating that association, rather than fusion, of vesicles is taking place. The

Fig. 4.27 DPH structure

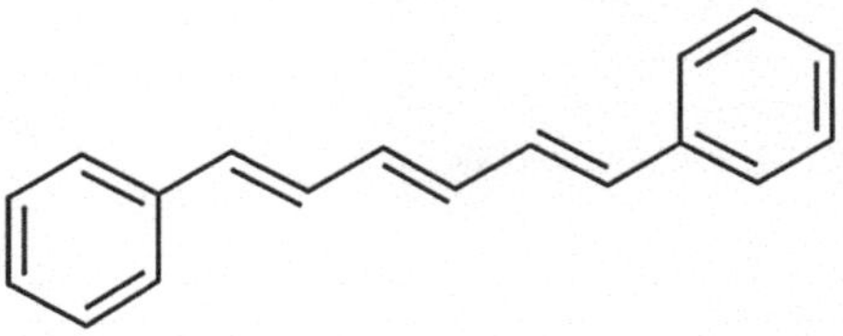

maximum in vesicle aggregation takes place at a peptide to lipid ratio that corresponds approximately to membrane neutrality. This behavior is consistent with what has been reported for other peptides [14, 83, 109, 161]. With neutral membranes, on the other hand, significant aggregation is induced only by P5Del, while the effect of P5 is negligible (Fig. 4.26).

4.2.6.3 Effect of the Peptide in the Thermotropic Phase Transition of Liposomes

For this experiment DPH-labeled DMPC vesicles in a 10 μM concentration were used, in the presence of 5 μM peptide. DPH (Fig. 4.27) inserts into the hydrophobic core of the membrane, essentially parallel to the lipid chains. For this reason, its fluorescence anisotropy reports on the membrane order and dynamics, and represents a measure of the membrane fluidity. Above the phase transition temperature, when the lipids are in the liquid-crystalline state, DPH anisotropy is very low, due to the fact that the probe is quite free to rotate around its axis [18]. Under the transition temperature, on the other hand, DPH anisotropy reaches a value around 0.35, because its mobility is largely inhibited when the lipids are in the ordered gel phase. The solution temperature was varied between 10 and 45 °C; DMPC vesicles showed the very nice liquid-to-gel phase transition at 24 °C. P5 had only a very slight effect on the membrane order and dynamics, even though it was present in a large excess (Fig. 4.28); P5Del, on the other hand, had a dramatic effect on the vesicles fluidity, with a strong broadening of the transition curve; in the presence of P5Del, indeed, the anisotropy values of DPH in the liquid and in the gel phase are much more similar than in the curve obtained without the peptide (Fig. 4.28).

The effects of the peptides on DPH anisotropy were investigated also in ePC vesicles at 25 °C; in these conditions, ePC lipids are in the physiological fluid state. A 10 μM solution of liposomes was titrated with increasing amounts of peptide, showing that also in this case P5Del has a stronger effect on DPH anisotropy, which increases in a concentration-related manner (Fig. 4.29).

Overall, these data concur to show that P5Del strongly perturbs both charged and neutral membranes, while P5 has a much higher activity on anionic vesicles than on neutral ones.

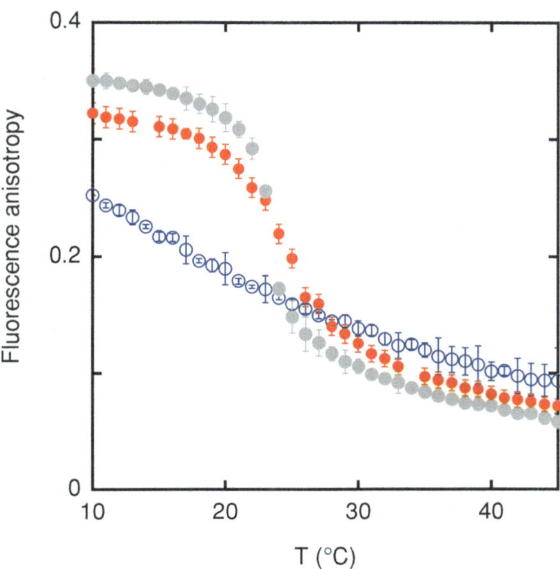

Fig. 4.28 Thermotropic gel to liquid crystalline phase transition in DMPC vesicles. [Lipid] = 10 μM; [peptide] = 5 μM. *Red*: P5, *blue*, P5Del, *grey*: no peptide [16]

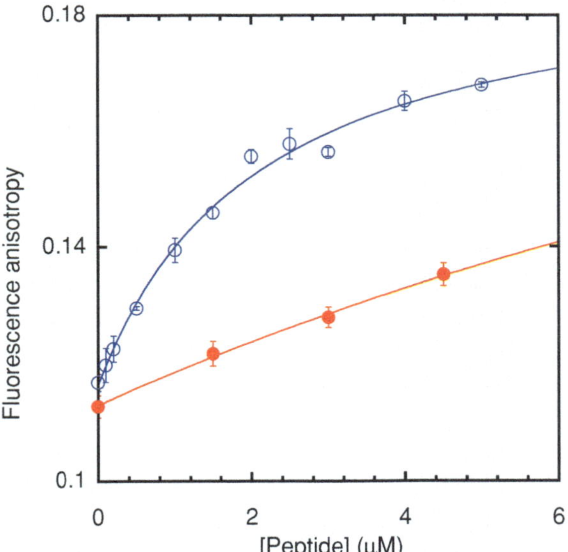

Fig. 4.29 peptide concentration dependence of DPH anisotropy in ePC vesicles. [Lipid] = 10 μM; T = 25 °C. *Grey* symbols: peptide-free vesicles; *red* symbols: vesicles with P5; *blue* symbols: vesicles with P5Del

4.2.7 Water-Membrane Partition Experiments

The different perturbing activity of P5 and P5Del on neutral membranes could be due to a different affinity for them, or to a different behavior of the peptides once bound to membranes with a similar affinity. To answer to this question, experiments were performed to investigate the peptide water-membrane partition equilibria.

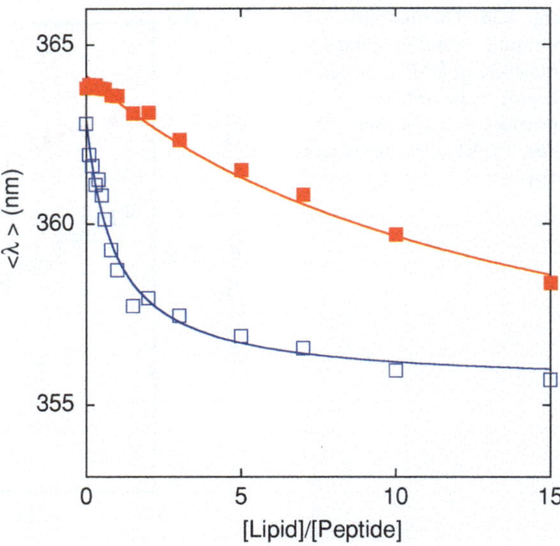

Fig. 4.30 Peptide water-membrane partition as measured from the blue-shift in the emission spectrum of the Trp residue in the presence of ePC/ePG (2:1) liposomes. [Peptide] = 1 μM. *Red symbols*: P5; *blue symbols*: P5Del [16]

The peptide intrinsic fluorescence, due to the presence of a Trp residue in the sequence, allowed us to perform fluorescence experiments.

4.2.7.1 Average Wavelength and Anisotropy Measurements

Peptide emission spectra were collected titrating a 1 μM solution of P5 or P5Del with increasing amounts of liposomes.

When the peptide is bound to the membrane, the Trp emission maximum is shifted towards lower wavelengths (i.e. the so-called *blue shift* happens), because the fluorophore is located in a less polar environment [18]. The extent of spectral changes was obtained calculating the average wavelength parameter

$$\langle\lambda\rangle = \sum_i \lambda_i I_i \Big/ \sum_i I_i$$

in the spectral interval between 320 and 420 nm.

In ePC/ePG liposomes, both P5 and P5Del exhibit a significant change in their emission spectrum, that is an evidence that the association is taking place (Fig. 4.30). In ePC/cholesterol liposomes, instead, only P5Del showed a spectral shift in its Trp maximum, while no significant change is shown by P5 (Fig. 4.31).

This result does not necessarily mean that P5 does not bind to neutral membranes, because there is the possibility for the peptide to attain a membrane-bound conformation which leaves the Trp residue exposed to the aqueous phase, so that its spectrum is not affected by the binding to the bilayer. To exclude this possibility, fluorescence anisotropy experiments have been performed. When free in

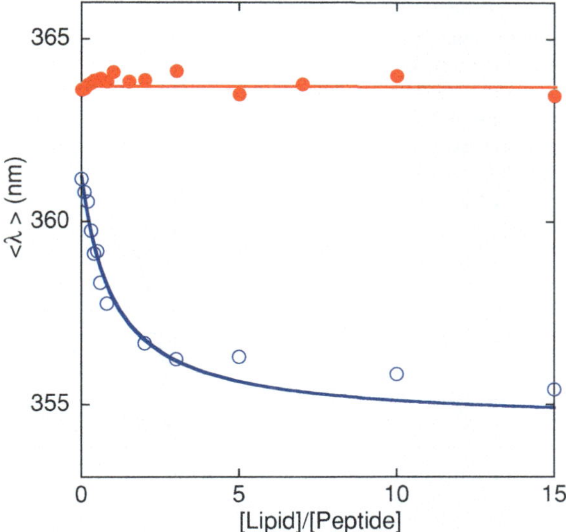

Fig. 4.31 Peptide water-membrane partition as measured from the blue-shift in the emission spectrum of the Trp residue in the presence of ePC/cholesterol (1:1) liposomes. [Peptide] = 1 μM. *Red symbols*: P5; *blue symbols*: P5Del [16]

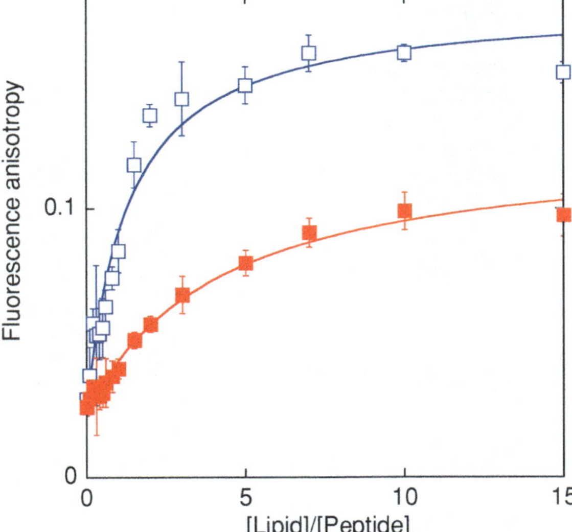

Fig. 4.32 Peptide water-membrane partition as measured from the peptide fluorescence anisotropy in the presence of ePC/ePG (2:1) liposomes. [Peptide] = 1 μM. *Red symbols*: P5; *blue symbols*: P5Del [16]

solution, the peptide rotational diffusion is very fast, and in this case Trp anisotropy is very low. When the peptide is associated to the membrane, on the other hand, its mobility is inhibited, and the anisotropy value increases [18]. These experiments were carried out with the same peptide and lipid concentrations used for the fluorescence spectra and showed very similar results: both peptides revealed affinity for negatively charged membranes (Fig. 4.32), while P5 does not bind to neutral bilayers (Fig. 4.33).

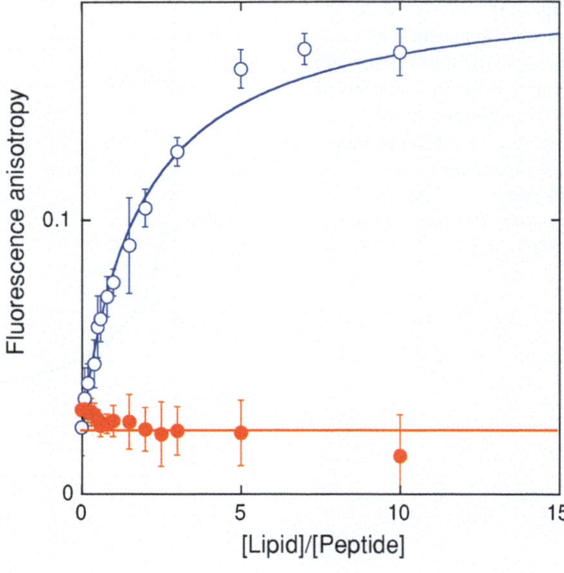

Fig. 4.33 Peptide water-membrane partition as measured from the peptide fluorescence anisotropy in the presence of ePC/cholesterol (1:1) liposomes. [Peptide] = 1 μM. *Red symbols*: P5; *blue symbols*: P5Del [16]

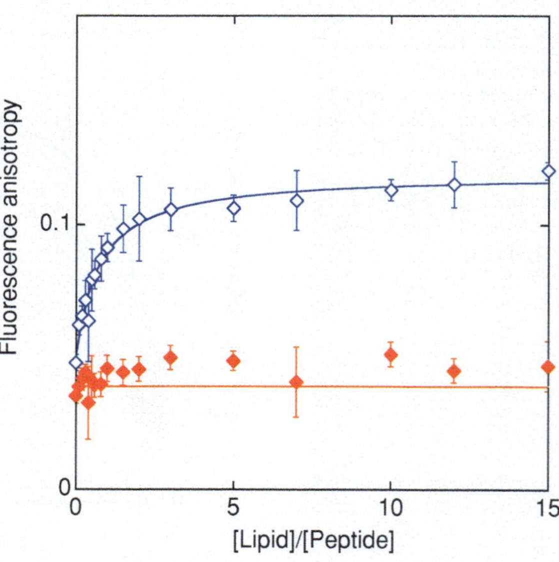

Fig. 4.34 Peptide water-membrane partition as measured from the peptide fluorescence anisotropy in the presence of ePC liposomes. [Peptide] = 1 μM. *Red symbols*: P5; *blue symbols*: P5Del [16]

Fluorescence anisotropy measurements were also performed with pure ePC vesicles, to find out if cholesterol plays a role also in the association phenomenon: P5, even in this case, did not show any evidence of binding (Fig. 4.34), proving that the lack of positive charge is the main factor inhibiting P5's binding, rather than cholesterol presence.

From these measurements it was also evident that the binding of P5 and P5Del with charged membranes occurs at relatively low peptide to lipid ratio, indicating a strong affinity of the two peptides for this kind of bilayers. The apparent partition

Table 4.4 Partition constants for P5 and P5Del in lipid vesicles of different composition

	P5	P5Del
ePC/ePG vesicles	$1.2\ 10^5$	$6.6\ 10^5$
ePC/chol vesicles	n.d.	$5.5\ 10^5$
ePC vesicles	n.d.	$1.3\ 10^6$

constants K_p were calculated from the fluorescence anisotropy data according to the following equation:

$$r_f + (r_b - r_f) * \frac{(K_p * [L])}{1 + (K_p * [L])}$$

where r_f and r_b are the Trp anisotropy values related to the free and membrane-bound peptide.

The resulting K_p are summarized in Table 4.4.

4.2.7.2 FRET Measurements-NBD

FRET (Sect. 2.1) of Chap. 2 can be used to determine the association of a peptide to a membrane. In the case of P5 and P5Del, the first fluorophore, the donor, is the Trp of our peptide. The second fluorophore, the acceptor, will be located within the membrane. FRET occurs, and can be detected, when the distance between the two fluorophores is comparable with the Förster radius. Under these conditions, the intensity of the donor fluorescence spectrum (Trp in this case) decreases, while the acceptor intensity increases.

A first set of experiments was performed using the fluorescent lipid NBD-PC. A peptide solution (1 μM) was titrated with ePC liposomes, labelled with NBD-PC (5 %) and the Trp and NBD fluorescence was detected. If FRET occurs, the intensity of NBD signal, located around 525 nm, increases. For P5Del, FRET occurs both in neutral and charged membranes, while for P5 a significant change in the NBD signal was observed only with ePC/ePG liposomes (Figs. 4.35 and 4.36).

4.2.7.3 FRET Measurements-DPH

Further FRET experiments were performed with DPH as acceptor; the Förster distance of the pair Trp-DPH is longer than that for Trp-NBD (Förster distances for the pairs are 40 and 16 Å, respectively) [71]. In this case it is possible to determine a peptide-membrane association even if the fluorophore of the peptide remains exposed to the aqueous phase, and thus the binding is not detectable with the experiments illustrated until now. This further experiment has been carried out to determine whether P5 actually does not bind to neutral membranes.

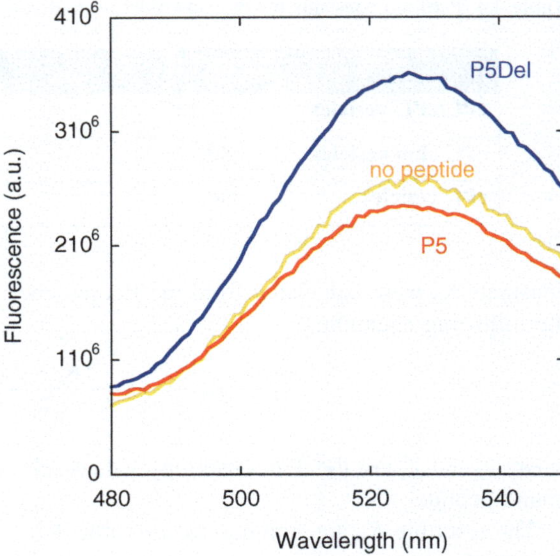

Fig. 4.35 NBD emission spectrum of labeled ePC/cholesterol liposomes in the absence (*yellow*) or in the presence of P5 (*red*) or P5Del (*blue*). [Peptide] = 1 μM, [Lipid] = 10 μM

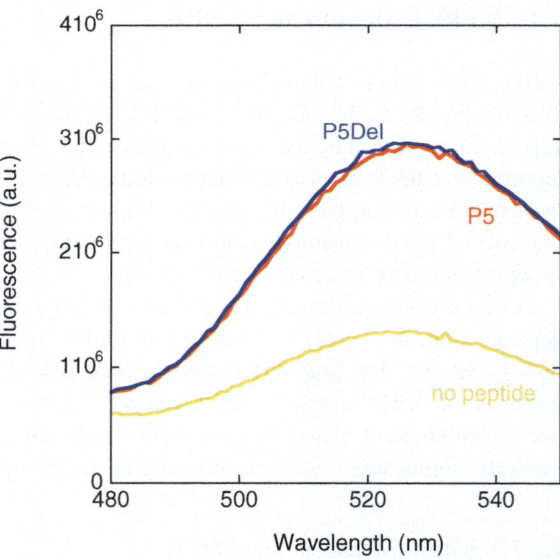

Fig. 4.36 NBD emission spectrum of labeled ePC/ePG liposomes in the absence (*yellow*) or in the presence of P5 (*red*) or P5Del (*blue*). [Peptide] = 1 μM, [Lipid] = 10 μM

In this case, DPH excitation spectra are reported: if FRET occurs, the Trp band appears in the spectrum. Once again, in neutral vesicles there is evidence of FRET only for P5Del, and for both peptides in neutral membranes (Figs. 4.37 and 4.38).

All the experiments indicate that P5 and P5Del exhibit a similar affinity for anionic model membranes. On the other hand, only P5Del significantly interacts with neutral membranes. Thus, the different peptides' activities are due to their different degree of interaction with lipid bilayers of different composition.

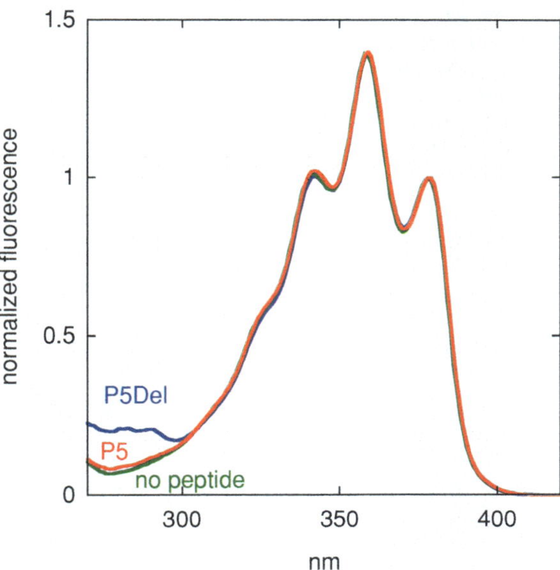

Fig. 4.37 DPH excitation spectra of ePC/cholesterol liposomes in the absence (*green*) or in the presence of P5 (*red*) or P5Del (*blue*). [Peptide] = 1 μM, [Lipid] = 10 μM

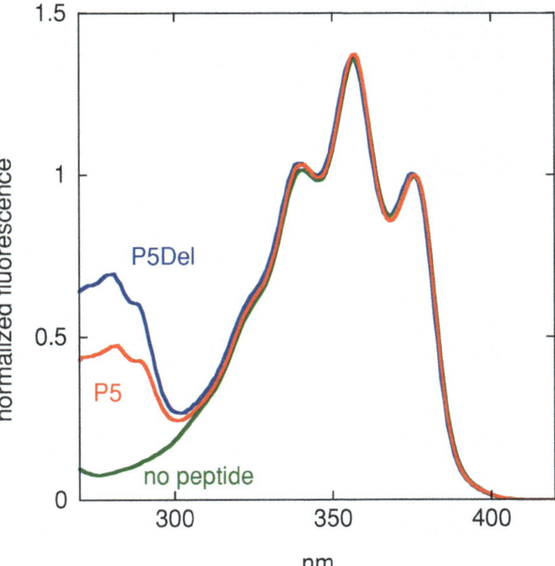

Fig. 4.38 DPH excitation spectra of ePC/ePG liposomes in the absence (*green*) or in the presence of P5 (*red*) or P5Del (*blue*). [Peptide] = 1 μM, [Lipid] = 10 μM

4.2.8 Peptide Structures in Water and Membranes

The affinity of the peptides for neutral membranes resulted dramatically affected by Pro removal. The explanation for this result is not obvious. Indeed, in practically all hydrophobicity scales, Pro contribution to the peptide preference for water

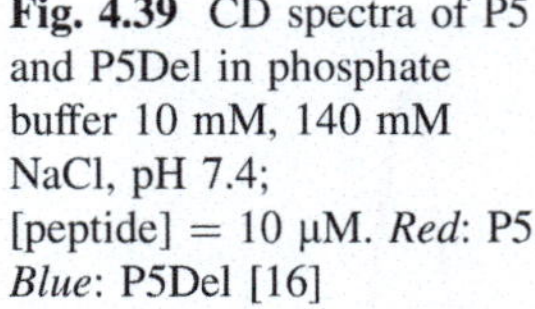

Fig. 4.39 CD spectra of P5 and P5Del in phosphate buffer 10 mM, 140 mM NaCl, pH 7.4; [peptide] = 10 μM. *Red*: P5; *Blue*: P5Del [16]

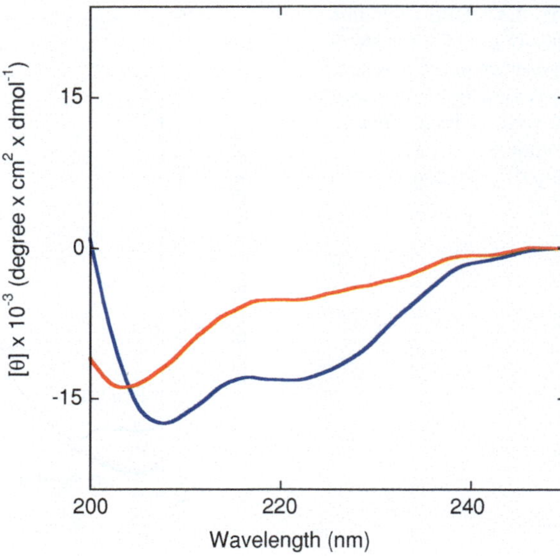

Fig. 4.40 CD spectra of P5 and P5Del in 50 % TFE. [Peptide] = 30 μM; *Red*: P5; *Blue*: P5Del [16]

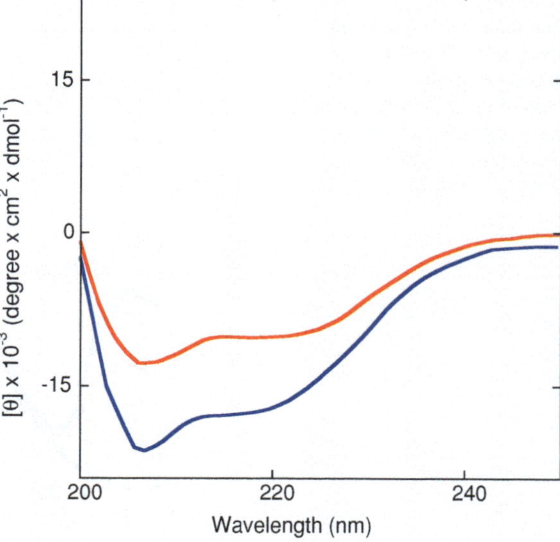

or apolar phases is substantially negligible. For instance, in the Wimley-White water-octanol partition scale [145] the Pro contribution to the free energy of partition is just 0.14 ± 0.11 kcal/mol, to be compared with 2.80 ± 0.11 kcal/mol for Lys, −1.25 ± 0.11 kcal/mol for Leu and −2.09 ± 0.11 kcal/mol for Trp. Therefore, Pro omission *per se* cannot be responsible for the different behavior of the two analogues. However, these calculations do not take into account peptide conformation. CD spectra showed that P5 and P5Del attain different conformations

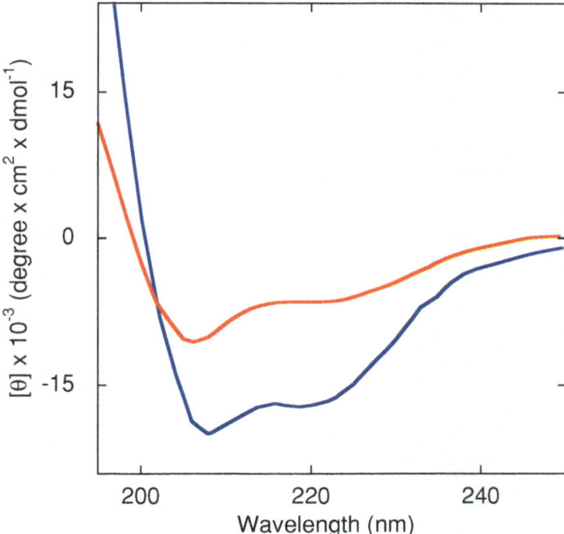

Fig. 4.41 CD spectra of P5 and P5Del in 30 mM SDS. [Peptide] = 30 = μM. *Red*: P5; *Blue*: P5Del [16]

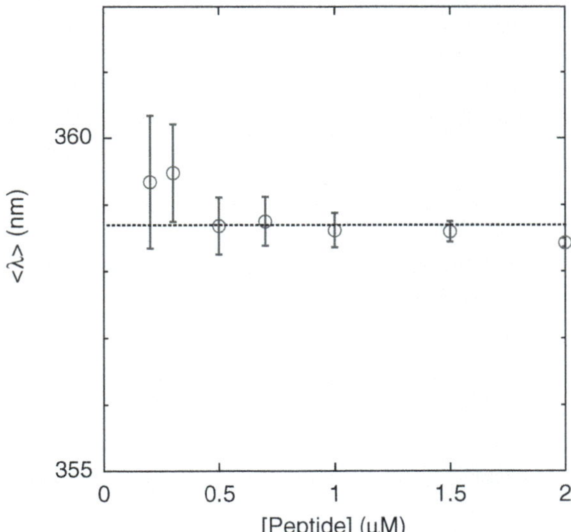

Fig. 4.42 Average wavelength of the peptide emission spectrum, as a function of peptide concentration. The error bars have been calculated repeating the experiment three times

in water: P5 is essentially unstructured, while P5Del features a significant amount of helical structure (Fig. 4.39). This result is consistent with the helix-breaking character of Pro in water. In a membrane-like environment, i.e. SDS or TFE both peptides showed a predominantly helical conformation (Figs. 4.40 and 4.41).

The relatively high helicity content in water of a quite short peptide such as P5Del might seem surprising. One possibility is that this structure is induced by aggregation. Incidentally, differences in aggregation have been sometimes invoked to explain differences in selectivity among AMP analogues [5, 66, 151, 152], although without providing a clear mechanistic explanation of the link between these two properties. To investigate this possibility, several concentration-dependent studies were performed. The fluorescence spectra did not show any significant variation (Fig. 4.42); thus, the hypothesis of aggregation can be excluded, at least in the low micromolar concentration range used in this study.

Probably, the lack of aggregation at relatively low concentrations is due to electrostatic repulsion caused by the high overall peptide net charge. At the same time, the helix stabilization of monomeric P5Del is provided by interactions between its hydrophobic residues, which in a helical conformation stack one on top of the other. Quantitative studies indicate that hydrophobic interactions between side chains located four residues apart in a helical peptide account for a stabilizing free energy difference that is 3–4 times the destabilizing effect due to the repulsion between two Lys residues [34]. Consequently, it is an established strategy in the design of helical peptides to integrate both charged and hydrophobic residues (but located on two different faces of the helix), to increase water solubility and secondary structure stability, respectively [2, 158].

The different conformations the two peptides attain in water have a dramatic effect on their effective hydrophobicity. A reliable measurement of this property is provided by the retention time in RP-HPLC. Several studies concur to show that this parameter is a good measure of peptide hydrophobicity, and takes into account also conformational effects [29, 47, 56, 82, 106, 153, 154, 159]. The observed retention time was 19 min for P5 and 25 min for P5Del. Based on the range of retention times observed for peptides of varying hydrophobicities [47, 106, 153], this is a very large difference, indicating that the P5Del is much more hydrophobic than P5.

4.2.9 Molecular Dynamics Simulations

In order to better elucidate the conformational effects of the Pro residue, MD simulations were performed for P5 and P5Del in the research group where this thesis was carried out. Five simulations, each 120 ns long, were performed for the two peptides in water, starting from a helical structure. In the initial stages of the simulations a conformational transition took place for both peptides, and equilibration was reached after simulation times ranging from about 5 to 70 ns, as judged from the matrixes of root mean square distances (RMSD) between trajectory structures [125]. The final, equilibrated 50 ns of the 5 trajectories of the two peptides were subjected to a cluster analysis, to find the predominant conformations, as reported in Table 4.5, and in Fig. 4.43. In the case of P5Del, all structures were very similar: one cluster (Fig. 4.43e) contained about 60 % of the structures, and the first 3 clusters contained almost 100 % of them (Fig. 4.43e–g). In all these structures P5Del maintained an essentially helical conformation, with some fraying at the termini. By contrast, simulations of P5 indicated a much higher conformational flexibility: in this case the first cluster represented only 40 % of the structures (Fig. 4.43a), and the first three only 76 % (Fig. 4.43a–c). In these

Table 4.5 Cluster analysis of the MD simulations in water

P5		P5Del	
Fraction of structures in the cluster	MLP	**Fraction of structures in the cluster**	MLP
39.0%	-27.42±0.05	**59.6%**	-26.52±0.04
19.8%	-27.55±0.05	**19.8%**	-28.12±0.06
17.2%	-27.81±0.07	**19.3%**	-26.98±0.07
Unclustered 24.0%	-28.33±0.08	**Unclustered 1.3%**	-26.8 ±0.4

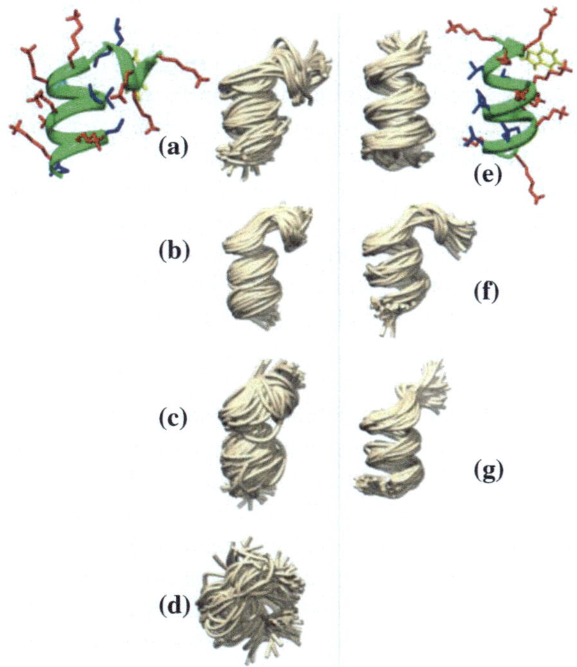

Fig. 4.43 Cluster analysis of MD simulations in water. Overlap of 50 structures, reported as a ribbon representation of the backbone, belonging to the clusters of the MD simulations of P5 (*left*, **a–d**) and P5Del (*right*, **e–g**) in water. The clusters are reported in order of relative population, from *top* to *bottom*. The last image for P5 (**d**) represents the unclustered structures. For the most populated cluster (*top*, **a** and **e**) the most representative structure (i.e. the structure with the smallest RMSD distance to the others) is also shown: the backbone is reported as a green ribbon, and side chains are shown as sticks (*red*: Lys; *blue*: Leu; *yellow*: Trp; *cyan*: Pro) [Reproduced from [16] with permission]

structures the peptide attained a compact conformation, allowed by a break in the helix caused by the Pro residue, where the hydrophobic residues were partially shielded from the solvent. In addition, the remaining 24 % of the P5 structures were not included in any cluster of significant size, and correspond to largely disordered conformations (Fig. 4.43d).

Conformational clusters obtained analyzing the C_α coordinates sampled each 250 ps in the last 50 ns of 5 independent 120 ns long simulations of P5 and P5Del in water. Only clusters containing more than 10 % structures are reported in the table. MLP values are reported with their standard errors.

These findings are in qualitative agreement with the CD spectra, which indicated a much higher helicity for P5Del, although the simulations overestimate the degree of structuring of P5, possibly as a result of biasing from the starting conformation. Anyway, the MD results provide an explanation for the different hydrophobicities and RP-HPLC retention times of the two analogues: the compact or bent conformations favored by the presence of Pro in P5 allow a partial shielding of the hydrophobic residues from the water phase. A rough but conformation-dependent assessment of peptide hydrophobicity in the simulated structures can be obtained with the molecular lipophilicity potential (MLP) [44]. The values reported in Table 4.5. confirm that the conformations sampled by P5 during the simulations are less hydrophobic than those of P5Del. This is particularly true for the unstructured P5 conformations, which, based on the CD data, are probably more populated than what our simulations show.

In principle, in addition to the peptide conformation in water, another property possibly causing the lack of P5 binding to neutral membranes could be its inability to attain a correct amphiphilic conformation in this membrane. Indeed, CD shows that P5 is largely helical once membrane bound, and, as shown in Fig. 4.19, a perfectly helical conformation would not give rise to the amphiphilic arrangement of the side chains needed for binding with high affinity to the bilayer. Obviously the free energy change associated with P5 binding to neutral membranes depends both on the starting and final states (P5 in water and in a PC bilayer, respectively). However, the latter condition is not attainable experimentally. Fortunately, MD simulations are particularly suited to investigate complex systems under experimentally inaccessible conditions. A simulation of P5 in the presence of POPC lipids was performed by using the so-called "minimum bias" approach [39, 99] in which the simulation starts from a random mixture of peptide, lipids and water (Fig. 4.44). In this way, a bilayer forms spontaneously during the simulation, but the peptide is free to find its minimum free energy position with respect to the lipids in the very dynamic environment of the initial stages of the simulation. As the starting structure for the peptide, a representative conformation of the most populated cluster obtained from the simulations in water was used. The lipid and water molecules present in the simulation were 128 and 7500, respectively, in a volume of about 390 nm^3 and this would correspond to a lipid concentration of about 0.55 M, i.e. probably high enough to ensure partition into the neutral membrane even for a peptide with a very low affinity such as P5. This is indeed what was observed: as shown in Fig. 4.44, P5 inserted into the membrane, lying

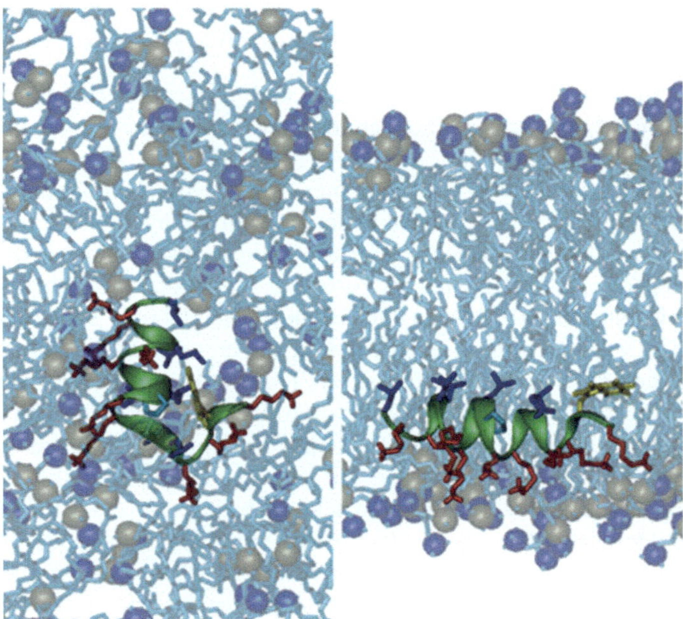

Fig. 4.44 MD simulations of P5 in the membrane. Structures at the beginning and end of the simulation of P5 are shown in the *left* and *right* panel, respectively. The peptide backbone is reported as a green ribbon, and side chains are shown as sticks (*red*: Lys; *blue*: Leu; *yellow*: Trp; *cyan*: Pro). For phospholipids, phosphorus and nitrogen atoms are represented as spheres (*grey* and *blue*, respectively), while the bonds as sticks. Water molecules are not shown, to simplify the image [Reproduced from [16] with permission]

parallel to the plane of the bilayer, just below the polar headgroups, in a position similar to that previously observed for other analogous peptides [99]. Notwithstanding the presence of the Pro residue, and in agreement with the CD findings, in the membrane P5 attained a helical conformation, although partially distorted with respect to the ideal geometry. The arrangement of intramolecular hydrogen-bonds was the canonical i to $i + 4$ for the N-terminal segment, then switched to i to $i + 5$ starting from the CO of residue 5, so that it was engaged in an H-bond notwithstanding the missing NH of Pro9. Eventually the H-bond pattern changed again to i to $i + 3$ at the N-terminus. This last shift left the CO groups of residues 12 and 13 lacking an intramolecular partner. However, their position in the membrane was rather superficial, thus allowing them to form H-bonds with water molecules, and to interact electrostatically with the headgroups of phospholipids. An important consequence of these deviations of the backbone dihedrals from the ideal helical angles (and probably the main driving force for these distortions of the helix) was the achievement of a perfect amphiphilic arrangement of the side-chains, with those of the hydrophobic residues embedded in the apolar core of the membrane. These findings suggest that the conformational properties of P5 in the membrane probably do not contribute to its low affinity for neutral bilayers.

4.2.10 Discussion

All the data clearly demonstrate a dramatic difference in the affinity of P5 and P5Del for neutral membranes. The main origin of this difference seems to be the peptides conformation in water.

In order to bind efficiently to membranes, AMPs have a sequence that allows them to attain an amphiphilic helical structure, with the hydrophobic residues on the same side of the helix. This alignment of hydrophobic residues is reminiscent of "Leu zippers" [1, 5, 101], and this structure is also present in these peptides. Hydrophobic interactions between Leu residues can stabilize helical conformations even in water, leaving a large exposed apolar surface, and favoring interaction with neutral membranes and toxicity. The data presented here indicate that the role of Pro residues in the sequence of AMPs is to destabilize these helical conformations, thus reducing the hydrophobic driving force for membrane binding. It is important to note that, on the other hand, peptide structure in the membrane is not very sensitive to amino acidic substitutions, and is usually helical [155]. Even peptides with a high content of Pro [99] or D-residues [97], or cyclized peptides [98] have been shown to be largely helical when bound to membranes, due to the lack of competition between intramolecular and intermolecular H-bonds, and to the free-energy cost of introducing a non-H-bonded amide bond in the membrane environment [145, 153, 154]. This is also what has been observed in the present study for the Pro-containing P5. Therefore, the most relevant factor in determining peptide toxicity is the helix stability in the water phase [138]. Destabilization of the helical structure in water reduces peptide affinity for neutral membranes by allowing a reduction in peptide hydrophobicity, due to the shielding of apolar residues, and also by introducing an entropic penalty for the membrane-binding process, due to the membrane-induced coil to helix transition. Indeed, a general correlation between peptide helicity and toxicity has been observed [25, 62, 81, 107, 138, 153–155] and some selective AMPs are imperfectly amphipathic [99, 107]. These findings were tentatively explained by proposing that helicity influences the peptide propensity to aggregate, or to interact with other components of the medium, which can reduce its activity by sequestering it [66]. However, aggregation was not relevant in the present instance. Therefore, it can be supposed, that a higher hydrophobicity, amphipathicity, affinity for neutral membranes, toxicity and tendency to aggregate are all possible consequences of a stable helical structure in water.

Several studies investigated the effect of substituting or inserting Pro residue on the activity and selectivity of specific AMPs, and provide further support to the present findings, and to the proposed interpretation. In most cases, substitution of Pro with another amino acid led to increased toxicity. For instance, this was observed for melittin [31], pardaxin [122, 134], tripticin [150], PMAP-23 [151, 152], temporin A [23] and model peptides [121]. By contrast, insertion of a Pro in peptides that did not contain it, usually led to the opposite effect, reducing peptide toxicity [12, 23, 124, 126, 157]. For instance, toxic peptides pisicidin [12], temporin

L [23] and the scorpion toxin IsCT [75] became selective once a Pro residue was inserted close to the center of their sequence. In some instances, it was shown that the Pro-induced changes in toxicity were mediated by an effect on the peptide affinity for neutral membranes [75, 124, 126, 151, 152]. Some exceptions to these general rules do exist: in some cases Pro insertions/substitution did not have an effect on hemolytic activity [13, 57], while in other cases they also affected significantly peptide activity against bacteria [3, 26, 110, 112, 157]. However, generally speaking, Pro insertion can be considered an effective method to reduce peptide toxicity, usually without perturbing significantly the antimicrobial activity. As a consequence of this, other similar approaches aiming at the destabilization of helical structures have been tested with success. Effects similar to those caused by Pro insertion/substitution were observed for Gly [24], another helix-destabilizing residue often found at the center of AMPs sequences. In addition, Shai and co-workers showed clearly that insertion of D-amino acids is a promising strategy to decrease peptide toxicity, and this approach was successfully applied to melittin, pardaxin, and model peptides [96, 102, 120]. Other investigators employed peptoid residues with the same goal [65]. However, peptides comprising D-residues or peptoids can be more immunogenic than sequences containing natural amino acids only, and are problematic for biotechnological production [11, 84]. Therefore Pro or Gly residues might be preferable to these alternatives.

In conclusion, these data provide a clear hypothesis for the role of Pro residues usually present in the sequence of AMPs: they are essential for peptide selectivity, by destabilizing helical conformations in the water phase, thus allowing conformations in which the apolar side-chains are partially buried, and reducing the hydrophobic driving force for binding to neutral membranes. These findings will be critical in the design of artificial molecules with a selectivity comparable to that of natural AMPs.

4.3 Tuning the Biological Activity

4.3.1 Cell-Penetrating Peptides

In this Chapter the possibility will be explored, to tune the biological activity of a peptide by limited modifications to its sequence. A case will be presented in which an AMP will be obtained from a cell-penetrating peptide (CPP).

CPPs are molecules endowed with the capability to cross eukaryotic cell membranes, and to deliver hydrophilic and macromolecular cargoes without causing membrane leakage [48, 72]. In some cases, it is possible for a CPP to deliver a molecule 100 fold bigger than its own weight [52, 79]. Therefore, they reveal a huge potential for gene and drug delivery applications [40], since in the last years protein or peptide-based drugs have been developed, but these molecules still feature important limitations, such as poor stability and low capability to reach their targets [48]: CPPs could represent a good solution to the problem of drug delivery.

CPPs have been firstly derived from proteins. In 1988 it was shown that the HIV-Tat protein is able to permeate the cell membrane, and to be internalized into the nucleus [43]; 10 years later, the Lebleu group demonstrated that that the minimal sequence needed for translocation was a 11-mer peptide (YGRKKRRQRRR) [144]. In 1991, Joliot et al. demonstrated that the *Drosophila Antennapedia* transcription factor [61] was also internalized by cells. The isolated domain, a 16-mer peptide (RQIKIYFQNRRMKWKK), was called penetratin, and it is now referred as pAntp [32, 79]. In the last twenty years, many chimeric or synthetic CPPs were obtained, also using unnatural amino acids to improve their systemic stability [48].

The mechanism of CPP internalization in many cases remains to be clarified. One of the most interesting hypotheses is that different mechanisms can compete for the uptake, also depending on the structural features of the peptide [48].

CPPs and AMPs have been designed, tested and analyzed as two distinct classes of peptides, due to their very different origin, and biological activities [51]. However, in many cases they share some common characteristics: many CPPs are short, amphiphilic and cationic, and they often attain a helical structure when associated to membranes, just like AMPs. The antibacterial activity of AMPs is strictly correlated with their structural features; thus, it is not a surprise, if some CPPs were shown also to be able to penetrate bacterial membranes [94, 100], and resulted to be, in this case, membranolytic.

For instance, the CPP pVEC was shown to kill bacteria by permeabilizing their membranes [100], and also transportan 10 (TP10) induces permeabilization of model membranes [149] and is bactericidal [94]. Both penetratin and Tat peptide were shown to be antimicrobial, with a MIC in the micromolar range. Similar results were obtained for the artificial peptide MAP [100]. On the other hand, while most AMPs exert their activity by perturbing the bacterial membrane, some of them are known to penetrate inside bacterial cells and to act on intracellular targets. The most prominent example of these AMPs is buforin [103], but many other exist [95]. However, even membrane-perturbing AMPs, such as magainin, were shown to penetrate into eukaryotic cells [133].

For these reasons, it is possible to ask which characteristics are responsible for different activities, or if there is really an effective separation between CPPs and AMPs. Since the activity of both CPPs and AMPs is strictly related to their structure, it is possible to hypothesize that it can be finely tuned with appropriate substitutions in the amino acidic sequence to switch between the two classes, i.e. it might be possible to obtain an AMP from a CPP, and *vice versa*. Such a result could be of great importance, because in this case the bactericidal activity could also be coupled with the ability to deliver a cargo in the cell compartment.

To clarify at least some of these points, the CPP called Pep-1 was chosen as a test case. Pep-1 (KETWWETWWTEWSQPKKKRKV), also called "Chariot" for its ability to transport cargoes inside cells, was designed by combining three domains [91]. The first is a hydrophobic, Trp-rich segment, KETWWETWWTEW, which is

Table 4.6 Comparison of Pep-1, Pep-1-K and melittin MICs against different bacterial strains. (MRSA = methicillin-resistant *S. aureus*. MDRPA = multidrug-resistant *P. aeruginosa*). From [160]

Bacterial strain	**MIC (μM)**		
	Pep-1	Pep-1-K	Melittin
Gram-(-) bacteria			
E.coli	32	2	2
S. typhimurium	>64	1	2
P. aeruginosa	>64	2	4
Gram-(+) bacteria			
B. Subtilis	8	2	2
S. aureus	64	2	1
S- epidermidis	64	1	1
Antibiotic-resistant bacteria			
MRSA1	>64	2	1
MRSA2	>64	2	2
MRSA3	>64	1	1
MDRPA1	32	8	16
MDRPA2	>64	8	8
MDRPA3	32	8	8

mainly responsible for peptide-protein interactions, that play a key role in the peptide carrier function. The second domain is constituted by a Pro-containing spacer, SQP, which gives flexibility to the peptide structure. Finally, a hydrophylic domain, crucial for the membrane translocation, is present (KKKRKV). The amphiphilic composition of Pep-1 allows it to interact with lipid bilayers [53], and to form non-covalent aggregates with cargo molecules, stabilized by both hydrophobic and electrostatic interactions [93].

Pep-1 exhibits a strong cell-permeating capability, but its translocation mechanism is still debated. The most promising hypothesis seems to involve a non-endocytotic pathway; furthermore, it has been proved that Pep-1 is able to perturb membrane permeability to translocate across the bilayer, and that a transmembrane potential is needed for activity [33, 50, 105, 160]. On the other hand, Pep-1 shows only a slight antimicrobial activity, at very high concentrations, for which it becomes also toxic for erythrocytes.

A Pep-1 analogue, called Pep-1-K, was designed, in which all negatively charged residues where replaced by Lys, with the aim to obtain a peptide more similar to AMPs, with a higher cationic charge [160]:

KKTWWKTWWTKWSQPKKKRKV

Indeed, Pep-1-K exhibited a strong antimicrobial activity, with MICs in the low μM range [160], as reported in Table 4.6. Pep-1-K has a bactericidal activity similar to that of melittin, without being toxic for erythrocytes.

Studies on *Staphylococcus aureus* cultures have shown that the antibiotic activity of Pep-1-K is related to its capability to cause the cell depolarization, reaching almost 100 % of depolarization at a peptide concentration of 64 μM [160].

These intriguing results raise a number of questions: do the sequence differences between Pep-1 and Pep-1-K influence only the antimicrobial activity of the two peptides or also their cell penetrating properties? What is the mechanism of the antibacterial activity of Pep-1-K? More specifically, is membrane depolarization caused directly by Pep-1-K association to the bacterial surface or is it just a consequence of some other effect of the peptide on the cell metabolism? Last, but not least, what are the causes of the switch in activity between Pep-1 and Pep-1-K?

To address these issues, the interaction of Pep-1 and Pep-1-K with cells and model membranes has been characterized.

4.3.2 Membrane-Perturbing Effects

4.3.2.1 Peptide-Induced Vesicle Leakage Assay

The membrane-perturbing activity of a peptide can be measured by its capability to cause the leakage from dye-loaded vesicles. For this purpose, ePC/ePG CF-loaded liposomes were used (200 μM, as described in Sect. 3.7) of Chap. 3 with the addition of several peptide concentrations.

The peptide was not able to cause the CF release, even at high concentration (Fig. 4.45).

4.3.2.2 Peptide-Induced Ion Leakage

The ability of Pep-1-K to cause the depolarization of bacterial membranes could be due to the formation of defects or pores in the membrane, too small to cause the leakage of a molecule like CF, but sufficiently large to allow the release of ions. To clarify this point, a sodium-sensing dye (SBFI, Fig. 4.46) was entrapped inside ePC/ePG lipid vesicles, which were then diluted them in a NaCl-containing buffer, while only KCl was present inside the vesicles. SBFI is a fluorescence probe, which excitation spectrum changes in the presence of sodium, due to the complex formation. The liposomes solution was titrated with increasing amounts of Pep-1-K, up to 10 μM.

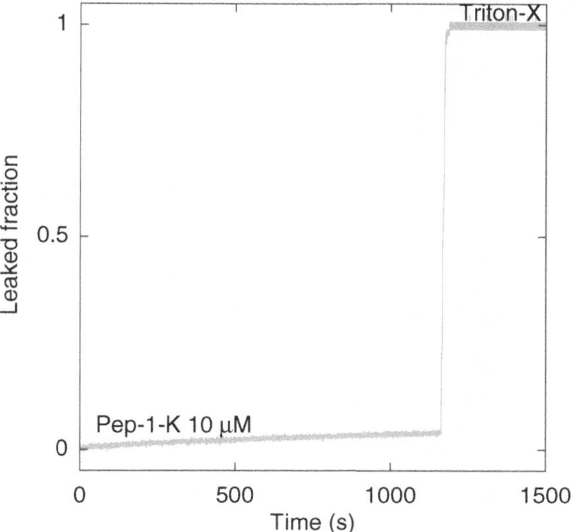

Fig. 4.45 Peptide-induced release from CF-loaded liposomes. Peptide concentration = 10 μM; lipid concentration: 200 μM [14]

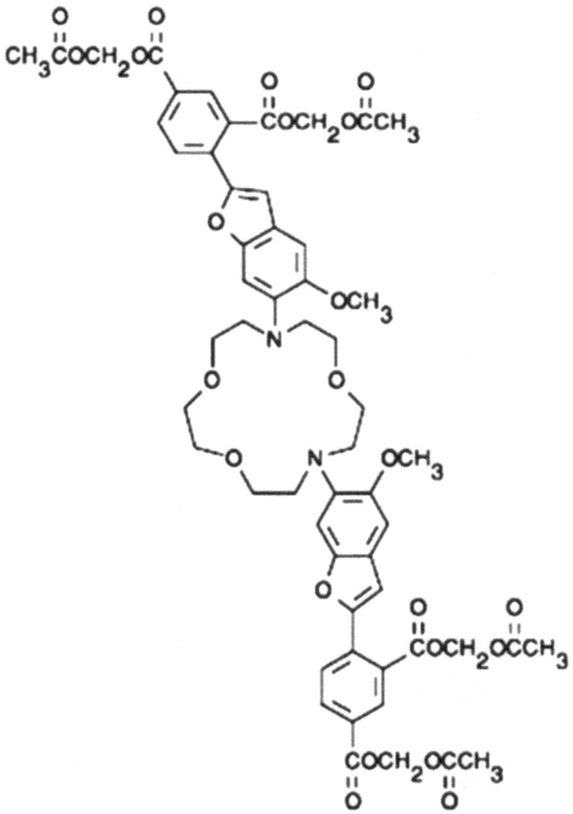

Fig. 4.46 SBFI probe structure

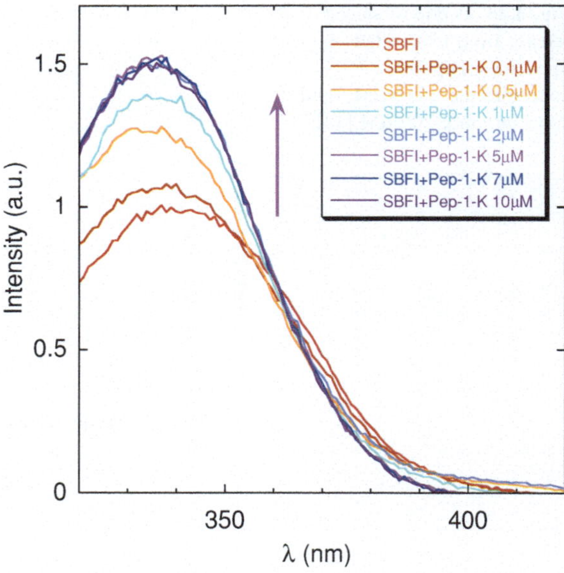

Fig. 4.47 Excitation spectra of SBFI in the presence of increasing amounts of Pep-1-K [14]

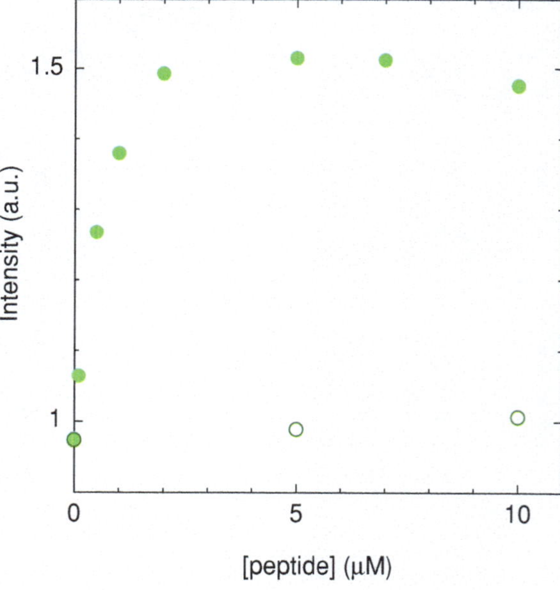

Fig. 4.48 Intensity increase of the SBFI spectrum at $\lambda = 335$ nm as a function of peptide concentration. Full circles: Pep-1-K; empty circles: Pep-1 [14]

The addition of Pep-1-K to a liposome solution caused Na^{+} entry into the vesicles, as shown by the change in the SBFI excitation spectrum. The intensity of the signal increased (Fig. 4.47), and the maximum was blue-shifted (Fig. 4.48). By contrast, Pep-1 was not able to cause such leakage at any concentration tested (up to 10 μM).

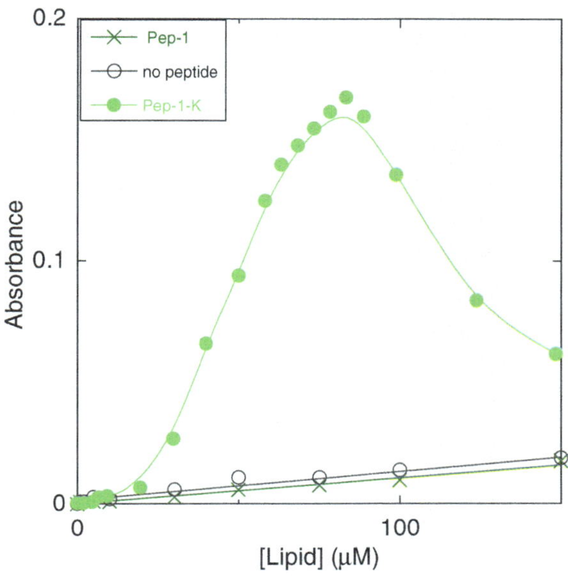

Fig. 4.49 Peptide-induced vesicle aggregation, as measured by the sample turbidity at 400 nm. Empty circles: peptide-free sample; full circles: Pep-1-K containing sample; crosses: Pep-1 containing sample. [Peptide] = 1 μM. ePC/ePG 2:1 (mol/mol) vesicles [14]

4.3.2.3 Peptide-Induced Vesicle Aggregation

Another important evidence of the membrane-perturbing activity of Pep-1-K is its ability to induce vesicles aggregation, which was detected by measuring the light scattering at 400 nm of a peptide solution, which was titrated with increasing amounts of ePC/ePG liposomes. Pep-1-K induced aggregation was reversible: the apparent absorbance reached a maximum around a lipid to peptide ratio of about 80, and then decreased (Fig. 4.49). No aggregation effect was detectable in the presence of Pep-1 (Fig. 4.49).

The phenomenon can be explained in terms of electrostatic attraction: when the peptide is bound to the anionic membranes, its positive charges progressively neutralize the negative charges of the lipids, favoring the formation of aggregates. When more liposomes are added, the peptide molecules are distributed over more vesicles, and therefore the total negative charge in each liposome increases again, causing a repulsion between vesicles, and a reduction in the aggregation. The reversibility of the increase in light scattering indicates that Pep-1-K causes vesicle aggregation rather than fusion, and therefore this process is probably not the origin of Pep-1-K induced ion leakage.

4.3.2.4 Effect of the Peptide in the Thermotropic Phase Transition of Liposomes

Another evidence of the peptide perturbing effect was obtained by measuring the phase transition temperature of the membrane. The experiment was carried out as described in Sect. 4.2.6.3).

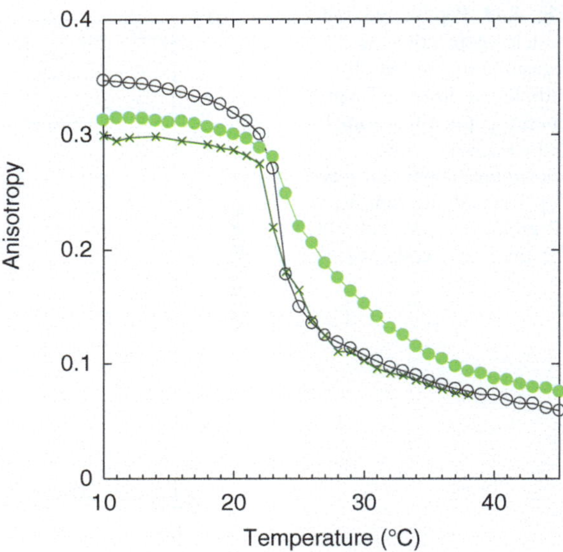

Fig. 4.50 Effect of peptide–membrane interaction on the thermotropic phase transition of DMPC/DMPG vesicles, as followed by the fluorescence anisotropy of DPH. Empty circles: peptide-free vesicles; filled circles: Pep-1-K associated vesicles; crosses: Pep-1 associated vesicles DMPC/DMPG 2:1 (mol/mol), DPH 1 %, [lipid] = 50 μM; [Peptides] = 10 μM [14]

As shown in Fig. 4.50, Pep-1-K binding significantly modifies the dynamics of a DMPC/DMPG bilayer both above and below the thermotropic phase transition, while no peptide-induced membrane perturbation was observed for Pep-1 in the physiologically relevant fluid state. Therefore peptide-induced membrane perturbation could be the basis of the Pep-1-K induced ion leakage.

4.3.3 Water-Membrane Peptide Partition

4.3.3.1 Steady-State Fluorescence

The origin of the increased membrane-perturbing activity of Pep-1-K, as compared to Pep-1, might be related to a different affinity of Pep-1-K for anionic lipid bilayers. To study the water-membrane partition equilibria, liposomes mimicking the composition of bacterial membranes (ePC/ePG in 2:1 molar ratio) were used. The intrinsic fluorescence of the peptide, which features 5 Trp residues in its structure, can be exploited to perform static or dynamic fluorescence studies. From the fluorescence spectra, it is possible to evaluate the affinity of the peptide towards lipid vesicles, from the calculation of the average wavelength for every lipid concentration, as already illustrated for P5. The average wavelength shift for both Pep-1-K and Pep.1 is shown in Fig. 4.51.

In the case of Pep-1-K and Pep-1, membrane binding caused a significant blue-shift in the emission spectrum of both peptides, but Pep-1-K exhibited a significantly higher affinity than Pep-1 for this kind of membranes. On the other hand, the spectral shift caused by membrane association was similar for both peptides, suggesting that they have a similar position and orientation in the membrane.

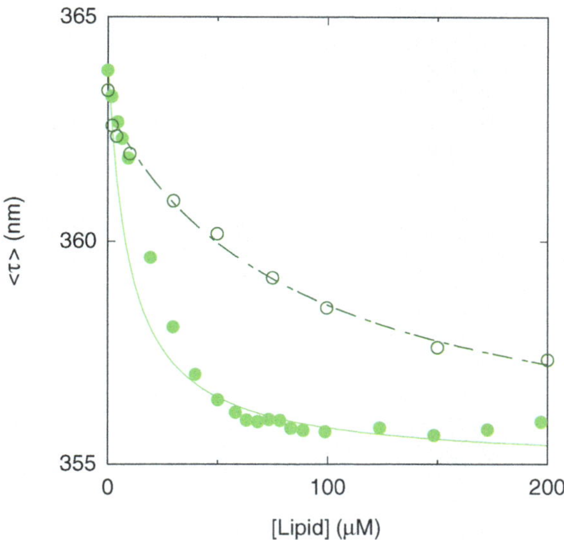

Fig. 4.51 Water to membrane partition of Pep-1 (*empty symbols*) and Pep-1-K (*filled symbols*), followed by the shift in the fluorescence emission spectrum. [Peptide] = 1 μM, ePC/ePG (2:1 mol/mol) vesicles ($\lambda_{exc.}$ = 280 nm ($\lambda_{em.}$ = 320–420 nm [14]

These findings indicate that the higher antibacterial activity of Pep-1-K is likely due, at least in part, to its higher membrane affinity.

4.3.3.2 Time-Resolved Fluorescence

The same peptide titration was performed also measuring the fluorescence decay lifetimes of Pep-1-K. The Trp decay profile in the presence of increasing amounts of liposomes was fitted with a triple-exponential function; from the data analysis it was possible to obtain the average lifetime value, $\langle\tau\rangle$ (Fig. 4.52).

From the dynamic fluorescence data the partition constant can be derived, according to the equation

$$\langle\tau\rangle = \langle\tau\rangle_{[L]=0} + \left(\langle\tau\rangle_{[L]=\infty} - \langle\tau\rangle_{[L]=0}\right) \frac{\frac{K_p}{[W]}[L]}{1 + \frac{K_p}{[W]}[L]}$$

where $\langle\tau\rangle_{[L\,=\,0]}$ e $\langle T\rangle_{[L\,=\,\infty]}$ are the values of the average lifetimes for the peptide alone and in the presence of lipids in saturation conditions, respectively. This equation has been used in the hypothesis that only two species are present, the free and the bound peptide. The partition constant derived from this expression resulted to be $K_p = 5.0 \times 10^6$.

The partition constant has been derived from time-resolved measurements and not from steady-state measurements because the former are not affected by experimental problems, like the tendency of the peptide to attach to the cuvette quartz walls.

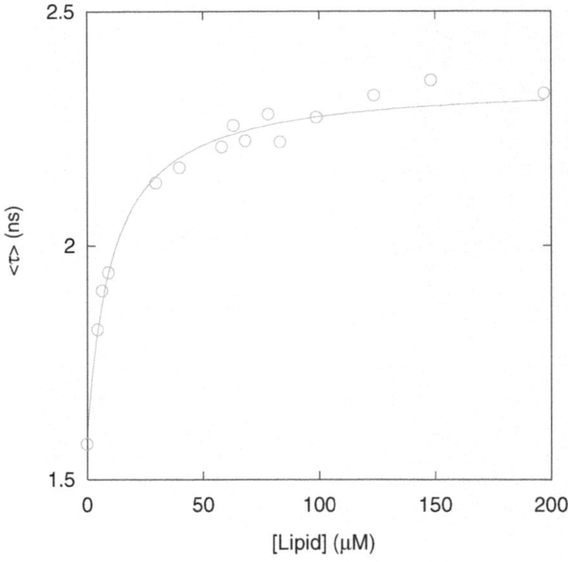

Fig. 4.52 ⟨τ⟩ variation of Pep-1-K as a function of lipid concentration. Experimental conditions: (λ_{ex} = 298 nm (λ_{em} = 343 nm; [Peptide] = 1 μM

4.3.3.3 Giant Unilamellar Vesicles

Direct observation of Pep-1-K association to membranes was possible using a fluorescein-labeled analogue and giant unilamellar vesicles (GUVs), labelled with Rho-PE (1 %). The peptide was added to a GUVs solution in 3 μM concentration.

Pep-1-K is located almost exclusively on the membrane surface, with negligible peptide fluorescence in the water phase both outside and inside GUVs, indicating that translocation does not occur (Fig. 4.53). This is probably due to the absence of a transmembrane potential, which was needed also for Pep-1 cell internalization [49].

Surprisingly, Pep-1-K did not associate homogeneously to all vesicles, and its fluorescence is concentrated only on certain GUVs, while does not appear on others. Anyway, this could be due to the impossibility of stirring the sample when the peptide was added in the observation chamber.

4.3.4 Peptide Location Inside the Bilayer

Depth-dependent quenching experiments were carried out to determine the position of Pep-1-K inside the lipid membrane. For this purpose, liposomes labelled with the nitroxyl group at different position of their acyl chain, or on the polar headgroups, were used. More precisely, 5-, 7-, 10-, 12-, 14-, 16-doxyl-PC and TEMPO-PC were used. Pep-1-K fluorescence spectra were collected in the presence of each type of liposomes, and also in a solution of unlabelled vesicles. All these experiments were performed at the same peptide and lipid concentration.

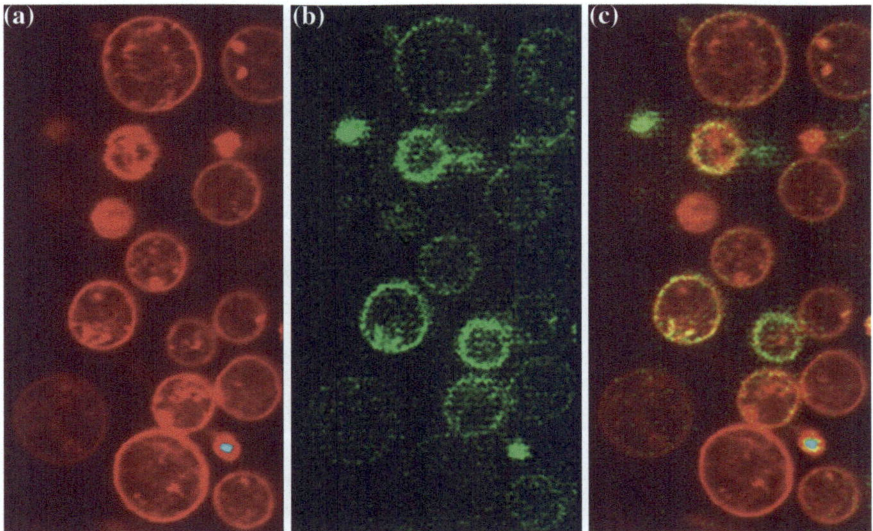

Fig. 4.53 Giant unilamellar vesicles images in the presence of Pep-1-K: panel *A*, vesicels fluorescence; panel *B*, peptide fluorescence; panel *C*, merge of the two signals. *A* and *B*. Image size 37.5 × 75 μm. GUV composition: ePC/ePG/Rho-PE 66:33:1 (molar ratios) [Reproduced from [14] with permission]

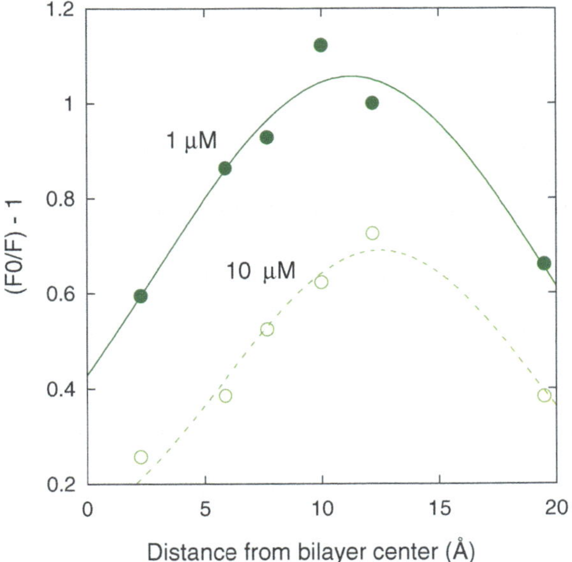

Fig. 4.54 Depth-dependent quenching experiment to determine the position of Pep-1-K in the membrane. [Peptide] = 1 μM (*full symbols*) or 10 μM (*empty symbols*), [lipid] = 200 μM. F and F_0 are the fluorescence intensities measured for the peptide associated to doxyl-labeled and unlabeled membranes, respectively. ($\lambda_{exc.}$ = 280 nm ($\lambda_{em.}$ = 320–420 nm. ePC/ePG (2:1 mol/mol) vesicles, doxyl labeled lipid content 7 % [14]

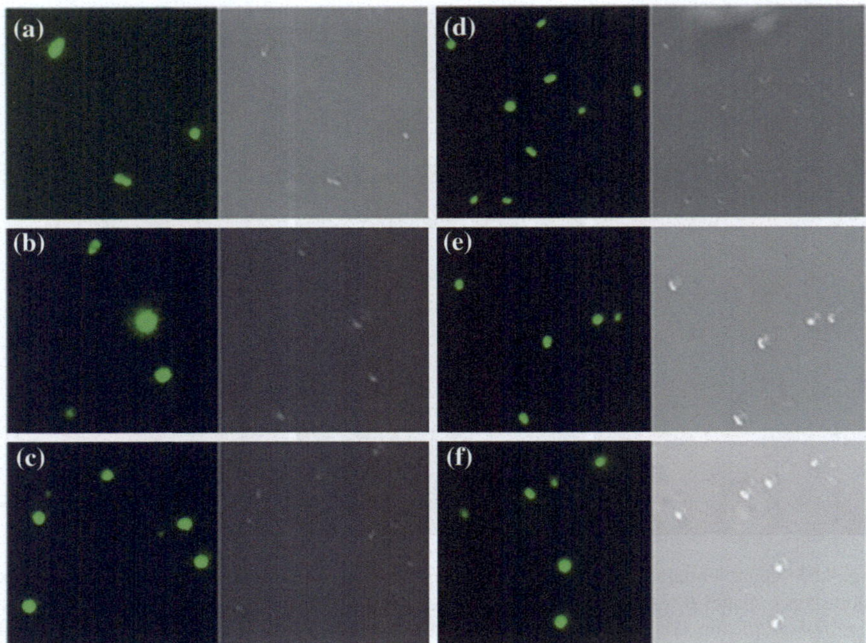

Fig. 4.55 Confocal laser-scanning and differential interference contrast (DIC) microscopy images (*left* and *right panels*, respectively) of *E.* coli treated with FITC-labeled peptides. Cells were treated with 10 μg/ml of FITC-labeled Pep-1 (*a*, *b* and *c*, representative images of different samples) or FITC-labeled Pep-1-K (*d*, *e* and *f*, representative images of different samples) [Reproduced from [14] with permission]

The nitroxyl moiety acts as a quencher of Trp fluorescence; thus, when the fluorophore is located at the same depth of the quencher group, its fluorescence will decrease. This is a short-range effect, so the quenching efficiency will rapidly decrease with the distance. From these experiments a quenching profile as a function of the distance from the bilayer center was obtained [99].

The quenching profile (Fig. 4.54) showed that Pep-1-K is located next to the membrane surface, right beneath the polar headgroups. The profile was rather well-defined, indicating that all the 5 Trp residues present in the peptide sequence are located at the same depth in the bilayer: the peptide lies parallel to the membrane surface, and its orientation does not change with peptide concentration (within the range investigated). This position and orientation suggest a mechanism of membrane perturbation that could be described according to the "carpet" model [119].

4.3.5 Cell-Penetrating Properties

To determine whether the cell-penetrating properties are retained by Pep-1-K, FITC-labeled analogues of Pep-1 and Pep-1-K were synthesized. The two peptides

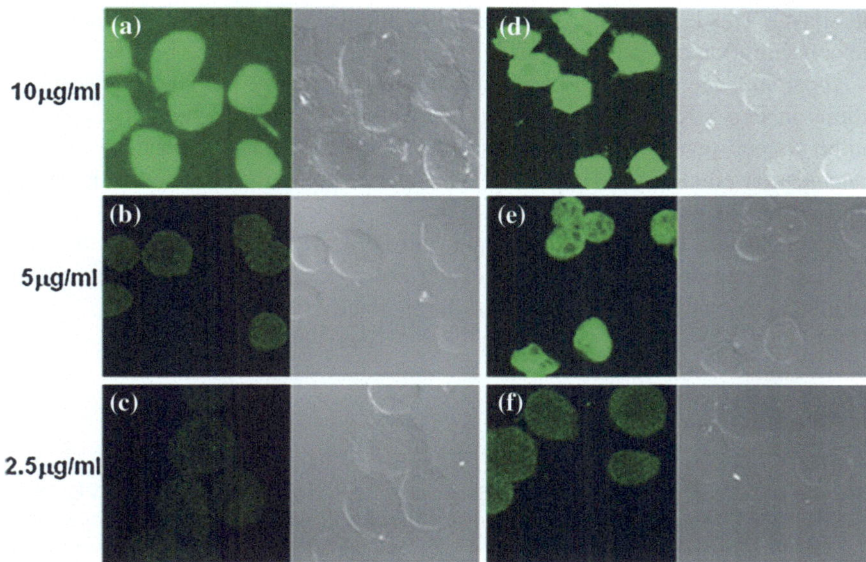

Fig. 4.56 Confocal laser-scanning and DIC microscopy images (*left* and *right panels*, respectively) of HeLa cells treated with FITC-labeled peptides. Cells were treated with FITC-labeled Pep-1 (*a*, *b* and *c*; 10, 5 and 2.5 μg/ml peptide concentration, respectively) and FITC-labeled Pep-1-K (*d*, *e* and *f*; 10, 5 and 2.5 μg/ml peptide concentration, respectively) [Reproduced from [14] with permission]

were then incubated with both bacterial and eukaryotic cells, and cell cultures were thereafter visualized by confocal microscopy. The collected images (Figs. 4.55 and 4.56) showed that both peptides were internalized in the cells, indicating that the Glu to Lys substitutions in Pep-1-K do not abolish the cell-penetrating activity.

4.3.6 Discussion

The data reported here indicate that the main difference between Pep-1 and Pep-1-K is in their relative affinities towards bacterial membranes: Pep-1-K binds to anionic bilayers more strongly, due to its higher cationic charge, and this appears to be the main reason for its strong bactericidal activity. This conclusion is in agreement with the correlation recently shown between water-membrane partition constants and MIC values of AMPs [87]. In the "carpet" model of peptide-induced membrane perturbation, AMPs need to reach a threshold of membrane-bound peptide concentration before they can cause the formation of defects or pores resulting in membrane leakage [119]. Therefore, it is evident that the higher the peptide affinity towards bacterial membranes, the lower is the concentration needed to reach this threshold. The reported data are consistent with a "carpet" model of membrane perturbation by Pep-1-K: it binds to the membrane surface, and perturbs the order of the bilayer. This leads to the leakage of ions, but not of larger molecules, at least in the concentration range investigated.

Apparently the change in membrane affinity caused by the Glu to Lys substitutions in Pep-1-K, while increasing its membrane-perturbing activity, does not inhibit its cell-penetrating properties. This is not surprising, since also the high translocation efficiency of Pep-1 itself has been shown to be linked to its strong affinity towards cellular membranes [53].

In conclusion, the example of Pep-1 and Pep-1-K clearly illustrates that cell-penetrating and antimicrobial peptides are not two separate classes, since subtle modifications can determine which of the two activities predominates.

References

1. Ahmad A, Azmi S, Srivastava RM, Srivastava S, Pandey BK, Saxena R, Bajpai VK, Ghosh JK (2009) Design of nontoxic analogues of cathelicidin-derived bovine antimicrobial peptide BMAP-27: the role of leucine as well as phenylalanine zipper sequences in determining its toxicity. Biochemistry 48:10905–10917
2. Albert JS, Hamilton AD (1995) Stabilization of helical domains in short peptides using hydrophobic interactions. Biochemistry 34:984–990
3. Andreu D, Merrifield RB (1985) N-Terminal analogues of cecropin a: synthesis, antibacterial activity, and conformational properties. Biochemistry 24:1683–1688
4. Andrews J (2001) Determination of minimum inhibitory concentrations. J Antimicrob Chemoth 48:5–16
5. Asthana N, Prasad Yadav S, Ghosh JK (2004) Dissection of antibacterial and toxic activity of melittin: a leucine zipper motif plays a crucial role in determining its hemolytic activity but not antibacterial activity. J Biol Chem 279:55042–55050
6. Auvin-Guette C, Rebuffat S, Prigent Y, Bodo B (1992) Trichogin A IV, an 11-residue lipopeptaibol from trichoderma longibrachiatum. J Am Chem Soc 114:2170–2172
7. Barlow DJ, Thornton JM (1988) Helix geometry in proteins. J Mol Biol 201:601–619
8. Baumann G, Mueller P (1974) A molecular model of membrane excitability. J Supramol Struct 2:538–557
9. Bechinger B, Skladnev DA, Ogrel A, Li X, Rogozhkina EV, Ovchinnikova TV, O'Neil JDJ, Raap J (2001) ^{15}N and ^{31}P solid-state NMR investigations on the orientation of zervamicin and alamethicin in phosphatidylcholine membranes. Biochemistry 40:9428–9437
10. Becucci L, Maran F, Guidelli R (2012) Probing membrane permeabilization by the antibiotic lipopeptaibol trichogin GAIV in a tethered bilayer lipid membrane. Biochim Biophys Acta 1818:1656–1662
11. Benkirane N, Friede M, Guichard G, Briand JP, Van Regenmortel MH, Muller S (1993) Antigenicity and immunogenicity of modified synthetic peptides containing D-amino acid residues. Antibodies to a D-enantiomer do recognize the parent L-hexapeptide and reciprocally. J Biol Chem 268:26279–26285
12. Beschiaschvili G, Seelig J (1990) Melittin binding to mixed phosphatidylglycerol/ phosphatidylcholine membranes. Biochemistry 29: 52–58 ((2007) Biochemistry 46(12):3653–3663
13. Blondelle SE, Houghten RA (1991) Hemolytic and antimicrobial activities of the twenty-four individual omission analogues of melittin. Biochemistry 30:4671–4678
14. Bobone S, Piazzon A, Orioni B, Pedersen J, Nan YH, Hahm KS, Shin SY, Stella L (2011) The thin line between cell-penetrating and antimicrobial peptides: the case of Pep-1 and Pep-1-K. J Pept Sci 17:335–341

15. Bobone S, Gerelli Y, De Zotti M, Bocchinfuso G, Farrotti A, Orioni B, Palleschi A, Sebastiani F, Latter E, Penfold J, Senesi R, Formaggio F, Toniolo C, Fragneto G, Stella L (2013a) Membrane thickness and the mechanism of action of the short peptaibol trichogin GAIV. Biochimica Biophysica Acta 1828:1013–1024
16. Bobone S, Bocchinfuso G, Park Y, Palleschi A, Hahm KS, Stella L (2013)b. The importance of being kinked: role of Pro residues in the selectivity of the helical antimicrobial peptide P5. J Peptide Sci 19:758–769
17. Bocchinfuso G, Palleschi A, Orioni B, Grande G, Formaggio F, Toniolo C, Park Y, Hahm KS, Stella L (2009) Different mechanisms of action of antimicrobial peptides: insights from fluorescence spectroscopy experiments and molecular dynamics simulations. J Pept Sci 15:550–558
18. Bocchinfuso G, Bobone S, Palleschi A, Stella L (2011) Fluorescence spectroscopy and molecular dynamics simulations in studies on the mechanism of membrane destabilization by antimicrobial peptides. Cell Mol Life Sci 68:2281–2301
19. Boheim G (1974) Statistical analysis of alamethicin channels in black lipid membranes. J Membr Biol 19:277–303
20. Bond PJ, Sansom MSP (2006) Insertion and assembly of membrane proteins via simulation. J Am Chem Soc 128:2697–2704
21. Brandl CJ, Deber CM (1986) Hypothesis about the function of membrane-buried proline residues in transport proteins. Proc Natl Acad Sci USA 4:917–921
22. Cafiso DS (1994) Alamethicin: a peptide model for voltage gating and protein-membrane interactions. Ann Rev Biophys Biomol Struct 23:141–165
23. Carotenuto A, Malfi S, Saviello MR, Campiglia P, Gomez-Monterrey I, Mangoni ML, Marcellini Hercolani Gaddi L, Novellino E, Grieco P (2008) A different molecular mechanism underlying antimicrobial and hemolytic actions of temporins A and L. J Med Chem 51:2354–2362
24. Chen HC, Brown JH, Morel JL, Huang CM (1988) Synthetic magainin analogues with improved antimicrobial activity. FEBS Lett 236:462–466
25. Chen Y, Mant CT, Farmer SW, Hancock REW, Vasil ML, Hodges RS (2005) Rational design of -helical antimicrobial peptides with enhanced activities and specificity/therapeutic Index. J Biol Chem 280:12316–12329
26. Chia BCS, Carver JA, Mulhern TD, Bowie JH (2000) Maculatin 1.1, an anti-microbial peptide from the Australian tree frog, Litoria genimaculata: solution structure and biological activity. Eur J Biochem 267:1889–2132
27. Cordes FS, Bright JN, Sansom MSP (2002) Proline-induced distortions of transmembrane helices. J Mol Biol 323:951–960
28. Daillant J (2005) Structure and fluctuations of a single floating lipid bilayer. Proc Natl Acad Sci USA 102:11639–11644
29. Dathe M, Meyer J, Beyermann M, Maul B, Hoischen C, Bienert M (2002) General aspects of peptide selectivity towards lipid bilayers and cell membranes studied by variation of the structural parameters of amphipathic helical model peptides. Biochim Biophys Acta 1558:171–186
30. Degenkolb T, Kirschbaum J, Brückner H (2007) New sequences, constituents and producers of peptaibiotics: an updated review. Chem Biodivers 4:1052–1067
31. Dempsey CE, Bazzo R, Harvey TS, Syperek I, Boheim G, Campbel ID (1991) Contribution of proline-14 to the structure and actions of melittin. FEBS Lett 281:240–244
32. Derossi D, Joliot A, Chassaing G, Prochiantz A (1994) The third helix of Antennapedia homeodomain translocates through biological membranes. J Biol Chem 269:10444–10450
33. Deshayes S, Plénat T, Charnet P, Divita G, Molle G, Heitz F (2006) Formation of transmembrane ionic channels of primary amphipathic cell-penetrating peptides. Consequences on the mechanism of cell penetration. Biochim Biophys Acta 1758:1846–1851
34. Doig AJ (2008) Stability and design of alpha-helical peptides. Progr Mol Biol Transl Sci 83:1–52

35. Duclohier H (2004) Helical kink and channel behaviour: a comparative study with the peptaibols alamethicin, trichotoxin and antiamoebin. Eur Biophys J 33:169–174
36. Duclohier H, Wróblewski H (2001) Voltage-dependent pore formation and antimicrobial activity by alamethicin and analogues. J Membr Biol 184:1–12
37. Duclohier H, Snook CF, Wallace BA (1998) Antiamoebin can function as a carrier or as a pore-forming peptaibol. Biochim Biophys Acta 1415:255–260
38. Epand RF, Epand RM, Monaco V, Stoia S, Formaggio F, Crisma M, Toniolo C (1999) The antimicrobial peptide trichogin and its interaction with phospholipid membranes. Eur J Biochem 266:1021–1028
39. Esteban-Martin S, Salgado J (2007) Self-assembling of peptide/membrane complexes by atomistic molecular dynamics simulations. Biophys J 92:903–912
40. Foerg C, Merkle HP (2008) On the biomedical promise of cell-penetrating peptides: limits versus prospects. J Pharm Sci 97:144–162
41. Formaggio F, Peggion C, Crisma M, Toniolo C (2001) Short-chain analogues of the lipopeptaibol antibiotic trichogin GA IV: conformational analysis and membrane modifying properties. J Chem Soc Perkin Trans 2:1372–1377
42. Fox RO, Richards FM (1982) A voltage-gated ion channel model inferred from the crystal structure of alamethicin at 1.5 Å resolution. Nature 300:325–330
43. Frankel AD, Pabo CO (1988) Cellular uptake of the Tat protein from human immunodeficiency virus. Cell 55:1189–1193
44. Gaillard P, Carrupt PA, Testa B, Boudon A (1994) Molecular lipophilicity potential, a tool in 3D QSAR: Method and applications. J Comp Aided Mol Des 8:83–96
45. Gatto E, Mazzuca C, Stella L, Venanzi M, Toniolo C, Pispisa B (2006) Effect of peptide lipidation on membrane perturbing activity: a comparative study on two trichogin analogues. J Phys Chem B 110:22813–22818
46. Grigoriev PA, Schlegel B, Kronen M, Berg A, Härtl A, Gräfe U (2003) Differences in membrane pore formation by peptaibols. J Pept Sci 9:763–768
47. Guo D, Mant CT, Taneja AK, Hodges RS (1986) Prediction of peptide retention times in reversed-phase high-performance liquid chromatography Correlation of observed and predicted peptide retention times factors and influencing the retention times of peptides. J Chromatogr 359:519–532
48. Heitz F, Morris MC, Divita G (2009) Twenty years of cell-penetrating peptides: from molecular mechanisms to therapeutics. Brit J Pharmacol 157:195–206
49. Henriques ST, Castanho MARB (2004) Consequences of nonlytic membrane perturbation to the translocation of the cell penetrating peptide Pep-1 in lipidic vesicles. Biochemistry 43:9716–9724
50. Henriques ST, Castanho MARB (2008) Translocation or membrane disintegration? Implication of peptide–membrane interactions in Pep-1 activity. J Pept Sci 14:482–487
51. Henriques ST, Melo MN, Castanho MARB (2006) Cell-penetrating peptides and antimicrobial peptides: how different are they? Biochem J 399:1–7
52. Henriques ST, Quintas A, Bagatolli LA, Homblé F, Castanho MARB (2007) Energy-independent translocation of cell-penetrating peptides occurs without formation of pores. A biophysical study with Pep-1. Mol Membr Biol 24:282–293
53. Henriques ST, Castanho MARB, Pattenden LK, Aguilar M (2010) Fast membrane association is a crucial factor in the peptide Pep-1 translocation mechanism: a kinetic study followed by surface plasmon resonance. Biopolymers (Pept Sci) 94:314–322
54. Heuber C, Formaggio F, Baldini C, Toniolo C, Müller K (2007) Multinuclear solid-state-NMR and FT-IR-absorption investigations on lipid/trichogin bilayers. Chem Biodivers 4:1200–1218
55. Hinderliter A, Biltonen RL, Almeida PFF (2004) Lipid modulation of protein-induced membrane domains as a mechanism for controlling signal transduction. Biochemistry 43:7102–7110
56. Houghten RA, Blondelle SE (1992) Design of model amphipathic peptides having potent antimicrobial activities. Biochemistry 31:12688–12694

57. Houghten RA, Blondelle SE (1994) Determination of the secondary structure of selected melittin analogues with different haemolytic activities. Biochem J 299:587–591
58. Huang HW (2009) Free Energies of molecular bound states in lipid bilayers: lethal concentrations of antimicrobial peptides. Biophys J 96:3263–3272
59. Hurley JH, Mason DA, Matthews BW (1992) Flexible-geometry conformational energy maps for the amino acid residue preceding a proline. Biopolymers 32:1443–1446
60. Ibrahim HR, Thomas U, Pellegrini A (2001) A helix-loop-helix peptide at the upper lip of the active site cleft of lysozyme confers potent antimicrobial activity with membrane permeabilization action. J Biol Chem 276:43767–43774
61. Joliot A, Pernelle C, Deagostini-Bazin H, Prochiantz A (1991) Antennapedia homeobox peptide regulates neural morphogenesis. Proc Natl Acad Sci USA 88:1864–1868
62. Khandelia H, Kaznessis YN (2006) Molecular dynamics investigation of the influence of anionic and zwitterionic interfaces on antimicrobial peptides' structure: Implications for peptide toxicity and activity. Peptides 27:1192–1200
63. Killian JA (1998) Hydrophobic mismatch between proteins and lipids in membranes. Biochim Biophys Acta 1376:401–416
64. Killian JA, Nyholm TKM (2006) Peptides in lipid bilayers: the power of simple models. Curr Op Struct Biol 16:473–479
65. Kim JK, Lee SA, Shin S, Lee JY, Jeong KW, Nan YH, Park YS, Shin SY, Kim Y (2010) Structural flexibility and the positive charges are the key factors in bacterial cell selectivity and membrane penetration of peptoid-substituted analog of Piscidin 1. Biochim Biophys Acta 1798:1913–1925
66. Kindrachuk J, Napper S (2010) Structure-activity relationships of multifunctional host defence peptides. Mini Rev Med Chem 10:596–614
67. Krishnakumar SS, London E (2007) Effect of sequence hydrophobicity and bilayer width upon the minimum length required for the formation of transmembrane helices in membranes. J Mol Biol 374:671–687
68. Kropacheva TN, Raap J (2002) Ion transport across a phospholipid membrane mediated by the peptide trichogin GA IV. Biochim Biophys Acta 1567:193–203
69. Kučerka N, Tristram-Nagle S, Nagle JF (2005) Structure of fully hydrated fluid phase lipid bilayers with monounsaturated chains. J Membr Biol 208:193–202
70. Kumar S, Bansa M (1998) Dissecting α-helices: Position-specific analysis of α-helices in globular proteins. Proteins: Struct Funct Bioinf 31:460–476
71. Lakowicz JR (2006) Principles of fluorescence spectroscopy, 3rd edn. Springer, New York
72. Langel Ü (2006) Handbook of cell penetrating peptides. CRC Press, Oxford
73. Langelaan DN, Wieczorek M, Blouin C, Rainey JK (2010) Improved helix and kink characterization in membrane proteins allows evaluation of kink sequence predictors. J Chem Inf Model 50:2213–2220
74. Laver DR (1994) The barrel-stave model as applied to alamethicin and its analogs reevaluated. Biophys J 66:355–359
75. Lee K, Shin SY, Kim K, Lim SS, Hahm KS, Kim Y (2004) Antibiotic activity and structural analysis of thescorpion-derived antimicrobial peptide IsCT and its analogs. Biochem Biophys Res Commun 323:712–719
76. Leitgeb B, Szekeres A, Manczinger L, Vágvölgyi C, Kredics L (2007) The history of alamethicin: a review of the most extensively studied peptaibol. Chem Biodivers 4:1027–1051
77. Liu Y, Engelman DM, Gerstein M (2002) Genomic analysis of membrane protein families: abundance and conserved motifs. Genome Biol 3:0054.1–0054.12
78. Lummis SCR, Beene DL, Lee LW, Lester HA, Broadhurst R, Dougherty DA (2005) Cis–trans isomerization at a proline opens the pore of a neurotransmitter-gated ion channel. Nature 438:248–252
79. Lundberg P, Langel Ü (2003) A brief introduction to cell-penetrating peptides (review). J Mol Recognit 16:227–233

80. Makino S, Kayahara T, Tashiro K, Takahashi M, Tsuji T, Shoji M (2001) Discovery of a novel serine protease inhibitor utilizing a structure-based and experimental selection of fragments technique. J Comp Aided Mol Des 15:553–559
81. Mangoni ML, Carotenuto A, Auriemma L, Saviello MR, Campiglia P, Gomez-Monterrey I, Malfi S, Marcellini L, Barra D, Novellino E, Grieco P (2011) Structure-activity relationship, conformational and biological studies of temporin L analogues. J Med Chem 54:1298–1307
82. Mant CT, Kovacs JM, Kim HM, Pollock DD, Hodges RS (2009) Intrinsic amino acid side-chain hydrophilicity/hydrophobicity coefficients determined by reversed-phase high-performance liquid chromatography of model peptides: comparison with other hydrophilicity/hydrophobicity scales. Biopolymers (Pept Sci) 92:573–595
83. Marquette A, Lorber B, Bechinger B (2010) Reversible liposome association induced by LAH4: a peptide with potent antimicrobial and nucleic acid transfection activities. Biophys J 98:2544–2553
84. Matsuzaki K (2009) Control of cell selectivity of antimicrobial peptides. Biochim Biophys Acta 1788:1687–1692
85. Mazzuca C, Stella L, Venanzi M, Formaggio F, Toniolo C, Pispisa B (2005) Mechanism of membrane activity of the antibiotic trichogin GA IV: a two-state transition controlled by peptide concentration. Biophys J 88:3411–3421
86. Mazzuca C, Orioni B, Coletta M, Formaggio F, Toniolo C, Maulucci G, De Spirito M, Pispisa B, Venanzi M, Stella L (2010) Fluctuations and the rate-limiting step of peptide-induced membrane leakage. Biophys J 99:1791–1800
87. Melo MN, Ferre R, Castanho MARB (2009) Antimicrobial peptides: linkong partition, activity and high-membrane-bound concentrations. Nat Rev Microbiol 7:245–250
88. Milov AD, Tsvetkov YD, Formaggio F, Crisma M, Toniolo C, Raap J (2000) Self-assembling properties of membrane-modifying peptides studied by PELDOR and CW-ESR spectroscopies. J Am Chem Soc 122:3843–3848
89. Milov AD, Tsvetkov YD, Formaggio F, Crisma M, Toniolo C, Raap J (2003) Self-assembling and membrane modifying properties of a lipopeptaibol studied by CW-ESR and PELDOR spectroscopies. J Pept Sci 9:690–700
90. Monaco V, Formaggio F, Crisma M, Toniolo C, Hanson P, Millhauser GL (1999) Orientation and immersion depth of a helical lipopeptaibol in membranes using TOAC as an ESR Probe. Biopolymers 50:239–253
91. Morris MC, Depollier J, Mery J, Heitz F, Divita G (2001) A peptide carrier for the delivery of biologically active proteins into mammalian cells. Nat Biotech 19:1173–1176
92. Mouritsen OG, Bloom M (1984) Mattress model of lipid-protein interactions in membranes. Biophys J 46:141–153
93. Munhoz-Morris MA, Heitz F, Divita G, Morris MC (2007) The peptide carrier Pep-1 forms biologically efficient nanoparticle complexes. Biochem Biophys Res Commun 355:877–882
94. Nekhotiaeva N, Elmquist A, Kuttuva Rajarao G, Hällbrink M, Langel Ü, Good L (2004) Cell entry and antimicrobial properties of eukaryotic cell-penetrating peptides. FASEB J 18:394–396
95. Nicolas P (2009) Multifunctional host-defense peptides: intracellular targeting antimicrobial peptides. FEBS J 276:6483–6496
96. Oren Z, Shai Y (1997) Selective lysis of bacteria but not mammalian cells by diastereomers of melittin: structure-function study. Biochemistry 36:1826–1835
97. Oren Z, Shai Y (2000) Cyclization of a cytolytic amphipathic r-helical peptide and its diastereomer:effect on structure, interaction with model membranes, and biological function. Biochemistry 39:6103–6114
98. Oren Z, Ramesh J, Avrahami D, Suryaprakash N, Shai Y, Jelinek R (2002) Structures and mode of membrane interaction of a short α helical lytic peptide and its diastereomer determined by NMR, FTIR, and fluorescence spectroscopy. Eur J Biochem 269:3869–3880
99. Orioni B, Bocchinfuso G, Kim JY, Palleschi A, Grande G, Bobone S, Park Y, Kim JI, Hahm KS, Stella L (2009) Membrane perturbation by the antimicrobial peptide PMAP-23: a fluorescence and molecular dynamics study. Biochim Biophys Acta 1788:1523–1533

100. Palm C, Netzereab S, Hällbrink M (2006) Quantitatively determined uptake of cell-penetrating peptides in non-mammalian cells with an evaluation of degradation and antimicrobial effects. Peptides 27:1710–1716
101. Pandey BK, Ahmad A, Asthana N, Azmi S, Srivastava RM, Srivastava S, Verma R, Vishwakarma AL, Ghosh JK (2010) Cell-selective lysis by novel analogues of melittin against human red blood cells and escherichia coli. Biochemistry 49:7920–7929
102. Papo N, Oren Z, Pag U, Sahl HS, Shai Y (2002) The consequence of sequence alteration of an amphipathic α-helical antimicrobial peptide and its diastereomers. J Biol Chem 277:33913–33921
103. Park, CB, Kim H. S., Kim, S.C (1998) Mechanism of action of the antimicrobial peptide buforin : Buforin kills microorganisms by penetrating the cell membrane and inhibiting cellular functions. Biochem Biophys Res Commun 244:253–257
104. Park Y, Lee DG, Jang SH, Woo ER, Jeong HG, Choi CH, Hahm KS (2003) A Leu–Lys-rich antimicrobial peptide: activity and mechanism. Biochim Biophys Acta 1645:172–182
105. Park N, Yamanaka K, Tran D, Chandrangsu P, Akers JC, de Leon JC, Morrissette NS, Selsted ME, Tan M (2009) The cell-penetrating peptide, Pep-1, has activity against intracellular chlamydial growth but not extracellular forms of Chlamydia trachomatis. J Antimicrob Chemother 63:115–123
106. Parker JMR, Guo D, Hodges RS (1986) New hydrophilicity scale derived from high-performance liquid chromatography peptide retention data: correlation of predicted surface residues with antigenicity and x-ray-derived accessible sites. Biochemistry 25:5425–5432
107. Pasupuleti M, Walse B, Svensson B, Malmsten M, Schmidtchen A (2008) Rational design of antimicrobial C3a analogues with enhanced effects against staphylococci using an integrated structure and function-based approach. Biochemistry 47:9057–9070
108. Peggion C, Formaggio F, Crisma M, Epand RF, Epand R, Toniolo C (2003) Trichogin: a paradigm for lipopeptaibols. J Pep Sci 9:679–689
109. Persson D, Thorén PEG, Bengt N (2001) Penetratin-induced aggregation and subsequent dissociation of negatively charged phospholipid vesicles. FEBS Lett 505:307–312
110. Pukala TL, Brinkworth CS, Carver JA, Bowie JH (2004) Investigating the importance of the flexible hinge in caerin 1.1: solution structures and activity of two synthetically modified caerin peptides. Biochemistry 43:937–944
111. Richardson JS, Richardson DC (1988) Amino acid preferences for specific locations at the ends of alpha helices. Science 240:1648–1652
112. Ringstad L, Schmidtchen A, Malmsten M (2010) Effects of single amino acid substitutions on peptide interaction with lipid membranes and bacteria–variants of GKE21, an internal sequence from human LL-37. Coll Surf A 354:65–71
113. Rodríguez A, Villegas E, Satake H, Possani LD, Corzo G (2011) Amino acid substitutions in an alpha-helical antimicrobial arachnid peptide affect its chemical properties and biological activity towards pathogenic bacteria but improves its therapeutic index. Amino Acid 40:61–68
114. Salnikov ES, Erilov DA, Milov AD, Tsvetkov YD, Peggion C, Formaggio F, Toniolo C, Raap J, Dzuba SA (2006) Location and aggregation of the spin-labeled peptide trichogin GAIV in a phospholipid membrane as revealed by pulsed EPR. Biophys J 91:1532–1540
115. Salnikov ES, Friedrich H, Li X, Bertani P, Reissmann S, Hertweck C, O'Neil JDJ, Raap J, Bechinger B (2009) Structure and alignment of the membrane-associated peptaibols ampullosporin A and alamethicin by oriented ^{15}N and ^{31}P solid-state NMR spectroscopy. Biophys J 96:86–100
116. Samatey FA, Xu C, Popot JL (1995) On the distribution of amino acid residues in transmembrane alpha-helix bundles. Proc Natl Acad Sci USA 92:4577–4581
117. Sansom MSP, Weinstein H (2000) Hinges, swivels and switches: the role of prolines in signalling via transmembrane α-helices. Trends Pharmn Sci 21:445–451
118. Senes A, Engel DE, Degrado WF (2004) Folding of helical membrane proteins: the role of polar, GxxxG-like and proline motifs. Curr Opin Struct Biol 14:465–479

119. Shai Y (2002) Mode of action of membrane active antimicrobial peptides. Biolpolymers (Pept Sci) 66:236–248
120. Shai Y, Oren Z (1996) Diastereomers of cytolysins, a novel class of potent antibacterial peptides. J Biol Chem 271:7305–7308
121. Shin SY, Yang ST, Park EJ, Eom SH, Song WK, Kim JI, Lee SH, Lee MK, Lee DG, Hahm KS, Kim Y (2001) Antibacterial, antitumor and hemolytic activities of α-helical antibiotic peptide, P18 and its analogs. J Pept Res 58:504–514
122. Sitaram N, Nagaraj R (1999) Interaction of antimicrobial peptides with biological and model membranes: structural and charge requirements for activity. Biochim Biophys Acta 1462:29–54
123. Smeazzetto S, De Zotti M, Moncelli MR (2011) A new approach to detect and study ion channel formation in microBLMs. Electrochem Comm 13:834–836
124. Song YM, Yang ST, Lim SS, Kim Y, Hahm KS, Kim JI, Shin SY (2004) Effects of L- or D-Pro incorporation into hydrophobic or hydrophilic helix face of amphipathic a-helical model peptide on structure and cell selectivity. Biochem Biophys Res Commun 314:615–621
125. Stella L, Melchionna S (1998) Equilibration and sampling in molecular dynamics simulations of biomolecules. J Chem Phys 109:10115–10118
126. Stella L, Mazzuca C, Venanzi M, Palleschi A, Didonè M, Formaggio F, Toniolo C, Pispisa B (2004) Aggregation and water-membrane partition as major determinants of the activity of the antibiotic peptide Trichogin GA IV. Biophys J 86:936–945
127. Stella L, Burattini M, Mazzuca C, Palleschi A, Venanzi M, Baldini C, Formaggio F, Toniolo C, Pispisa B (2007) Alamethicin interaction with lipid membranes: a spectroscopic study on synthetic analogues. Chem Biodivers 4:1299–1312
128. Subasinghage AP, Conlon JM, Hewage CM (2010) Development of potent anti-infective agents from Silurana tropicalis: Conformational analysis of the amphipathic, alpha-helical antimicrobial peptide XT-7 and its non-haemolytic analogue [G4 K]XT-7. Biochim Biophys Acta 1804:1020–1028
129. Suh JY, Lee YT, Park CB, Lee KH, Kim SC, Choi BS (1999) Structural and functional implications of a proline residue in the antimicrobial peptide gaegurin. Eur J Biochem 2:665–674
130. Syryamina VN, Isaev NP, Peggion C, Formaggio F, Toniolo C, Raap J, Dzuba SA (2010) Small-amplitude backbone motions of the spin-labeled lipopeptide trichogin GAIV in a lipid membrane as revealed by electron spin echo. J Phys Chem B 114:12277–12283
131. Syryamina VN, De Zotti M, Peggion C, Formaggio F, Toniolo C, Raap J, Dzuba SA (2012) A molecular view on the role of cholesterol upon membrane insertion, aggregation, and water accessibility of the antibiotic lipopeptide trichogin GAIV as revealed by EPR. J Phys Chem B 116:5653–5660
132. Tack BF, Sawai MV, Kearney WR, Robertson AD, Sherman MA, Wang W, Hong T, Boo LM, Wu H, Waring AJ, Lehrer RI (2002) SMAP-29 has two LPS-binding sites and a central hinge. Eur J Biochem 269:1181–1189
133. Takeshima K, Chikushi A, Lee KK, Yonehara S, Matsuzaki K (2003) Translocation of analogues of the antimicrobial peptides magainin and buforin across human cell membranes. J Biol Chem 278:1310–1315
134. Thennarasu S, Nagaraj R (1996) Specific antimicrobial and hemolytic activities of 18-residue peptides derived from the amino terminal region of the toxin pardaxin. Protein Eng 9:1219–1224
135. Thomas R, Vostrikov VV, Greathouse DV, Koeppe RE (2009) Influence of proline upon the folding and geometry of the WALP19 transmembrane peptide. Biochemistry 48:11883–11891
136. Toniolo C, Brückner H (eds) (2009) Peptaibiotics. Wiley-VCH, Weinheim
137. Toniolo C, Peggion C, Crisma M, Formaggio F, Shui X, Eggleston DS (1994) Structure determination of racemic trigogin A IV using centrosymmetric crystals. Nature Struct Biol 1:908–914

138. Tossi A, Sandri L, Giangaspero A (2000) Amphipathic, α-helical antimicrobial peptides. Biopolymers (Pept Sci) 55:4–30
139. Ulmschneider MB, Sansom MSP (2001) Amino acid distributions in integral membrane protein structures. Biochim Biophys Acta 1512:1–14
140. Venanzi M, Gatto E, Bocchinfuso G, Palleschi A, Stella L, Baldini C, Formaggio F, Toniolo C (2006)a Peptide folding dynamics: a time-resolved study from the nanosecond to the microsecond time regime. J Phys Chem B 110:22834–22841
141. Venanzi M, Gatto E, Bocchinfuso G, Palleschi A, Stella L, Formaggio, F, Toniolo C (2006)b Dynamics of formation of a helix-turn-helix structure in a membrane-active peptide: a time-resolved spectroscopic study. Chem Bio Chem 7:43–45
142. Venanzi M, Bocchinfuso G, Gatto E, Palleschi A, Stella L, Formaggio F, Toniolo C (2009) Metal binding properties of fluorescent analogues of Trichogin GA IV: a conformational study by time-resolved spectroscopy and molecular mechanics investigations. Chem Bio Chem 10:91–97
143. Vijayan K, Discher DE, Lal J, Janmey P, Goulian M (2005) Interactions of membrane-active peptides with thick, neutral, nonzwitterionic bilayers. J Phys Chem B 109:14356–14364
144. Vivés E, Brodin P, Lebleu B (1997) A truncated HIV-1 Tat protein basic domain rapidly translocates through the plasma membrane and accumulates in the cell nucleus. J Biol Chem 272:16010–16017
145. White SH, Wimley WC (1999) Membrane protein folding and stability: physical principles. Annu Rev Biophys Biomol Struct 28:319–365
146. Williams KA, Deber CM (1991) Proline residues in transmembrane helixes: structural or dynamic role? Biochemistry 30:8919–8923
147. Wong HT, Bowie JH, Carver JA (1997) The solution structure and activity of Caerin 1.1, an antimicrobial peptide from the australian green tree frog, Litoria Splendida. Eur J Biochem 247:545–557
148. Xiao Y, Herrera AI, Bommineni YR, Soulages JL, Prakash O, Zhang G (2009) The central kink region of Fowlicidin-2, an α-helical host defense peptide, is critically involved in bacterial killing and endotoxin neutralization. J Innate Immun 1:268–280
149. Yandek LE, Pokorny A, Floren A, Knoelke K, Langel U, Almeida PFF (2007) Mechanism of the cell-penetrating peptide transportan 10 permeation of lipid bilayers. Biophys J 92:2434–2444
150. Yang ST, Shin SY, Kim YC, Kim Y, Hahm KS, Kim JI (2002) Conformation-dependent antibiotic activity of tritrpticin, a cathelicidin-derived antimicrobial peptide. Biochem Biophys Res Comm 296:1044–1050
151. Yang ST, Jeon JH, Kim Y, Shin SY, Hahm KS, Kim JI (2006a) Possible role of a PXXP central hinge in the antibacterial activity and membrane interaction of PMAP-23, a member of cathelicidin family. Biochemistry 45:1775–1784
152. Yang ST, Lee JY, Kim HJ, Eu YJ, Shin SY, Hahm K-S, Kim JI (2006b) Contribution of a central proline in model amphipathic a-helical peptides to self-association, interaction with phospholipids, and antimicrobial mode of action. FEBS J 273:4040–4054
153. Zelezetsky I, Pacor S, Pag U, Papo N, Shai Y, Sahl HG, Tossi A (2005a) Controlled alteration of the shape and conformational stability of a-helical cell-lytic peptides: effect on mode of action and cell specificity. Biochem J 390:177–188
154. Zelezetsky I, Pag U, Sahl HG, Tossi A (2005b) Tuning the biological properties of amphipathic -helical antimicrobial peptides: rational use of minimal amino acid substitutions. Peptides 26:2368–2376
155. Zelezetsky I, Pontillo A, Puzzi L, Antcheva N, Segat L, Pacor S, Crovella S, Tossi A (2006) Evolution of the primate cathelicidin: correlation between structural variations and antimicrobial activity. J Biol Chem 281:19861–19871
156. Zhang L, Gazit EA, Boman HG, Shai Y (1995) Interaction of the mammalian antimicrobial peptide cecropin P1 with phospholipid vesicles. Biochemistry 34:11479–11488

157. Zhang L, Benz R, Hancock REW (1999) Influence of proline residues on the antibacterial and synergistic activities of α-helical peptides. Biochemistry 38:8102–8111
158. Zhang Y, Lu H, Lin Y, Cheng J (2011) Water-soluble polypeptides with elongated, charged side chains adopt ultrastable helical conformations. Macromolecules 44:6641–6644
159. Zhou NE, Mant CT, Hodges RS (1990) Effect of preferred binding domains on peptide retention behavior in reversed-phase chromatography: amphipathic alpha-helices. Pept Res 3:8–20
160. Zhu WL, Lan H, Park IS, Kim JI, Jin HZ, Hahm K-S, Shin SY (2006) Design and mechanism of action of a novel bacteria-selective antimicrobial peptide from the cell-penetrating peptide Pep-1. Biochim Biophys Res Commun 349:769–774
161. Ziegler A, Blatter XL, Seelig A, Seelig J (2003) Protein transduction domains of HIV-1 and SIV TAT interact with charged lipid vesicles. Binding mechanism and thermodynamic analysis. Biochemistry 42:9185–9194

Part II
Hydrogel Nanoparticles for Enzyme-Based Therapies

Chapter 5
Introduction

5.1 Enzymes as Drugs

Enzymes and proteins represent a very interesting class of therapeutics. They are widely used in different fields, i.e. as catalyzers in industrial processes (mainly drug synthesis) [43], as reagents in clinical and immunological chemistry [15], or in biosensors production [38], but their most important potential application is obviously related to therapeutic purposes [12, 23, 39, 58, 59].

Several diseases are related to the lack or to the defective activity of a particular enzyme. Among them, an example is represented by the so-called lysosomial storage diseases (LSD). Lysosomes are organelles involved in the storage and degradation of metabolites. LSD involve the lack of an enzyme in the lysosome compartment, leading to an accumulation of metabolites within cells, and include about 40 relatively rare different pathologies. Each of these diseases is related to the lack of a different enzyme [11, 23].

The study of LSDs has been fundamental for the development of enzyme replacement therapy (ERT) and in some cases, it has been possible to successfully administrate the lacking enzyme [39, 52, 53, 56]. In particular, the Gaucher disease Type 1 was the first intracellular enzymatic disease to be successfully treated using ERT in the 1990s. In the 2000s, it was followed by other LSDs, like mucopolysaccharidosis Type 1 and Pompe syndromes. In particular, the drug for the treatment of Gaucher disease, Ceredase®, obtained with recombinant DNA technology [57], was the first example of a successful ERT with an exogenous enzyme. These positive results have opened new frontiers in the development of ERT, supported by the molecular revolution of biotechnologies, which allowed the production of fully functional enzymes in large quantities from different species [23].

Nowadays several enzyme-based pharmacological formulations are available. For example, Adagen® is used to treat patients with a severe combined immunodeficiency disease, due to the lack of the adenosine deaminase enzyme. This example of ERT is particularly important, because it was the first enzyme approved under the so-called Orphan Drug Act, a fund created with the intent to encourage the research for treatment of diseases which affect a relatively small

S. Bobone, *Peptide and Protein Interaction with Membrane Systems*,
Springer Theses, DOI: 10.1007/978-3-319-06434-5_5,

number of patients, and that are often excluded from the research plans of big pharmaceutical companies. All the reported examples involve the intravenous administration of an enzyme of human origin, but there are also several cases of oral formulations, mainly targeted to gastrointestinal diseases.

On the other hand, ERT is still affected by severe problems, such as the immunologic response: several clinical studies [23] showed that enzyme administration often causes the development of antienzyme antibodies. Although this does not necessarily affect the therapeutic efficacy of protein-based drugs, this serious side-effect must be taken into account. Stability and administration of protein drugs are other complex challenges.

5.2 Stability and Administration

The activity of proteins is affected by different factors, such as pH, temperature, and by the maintenance of the correct structure. Therefore proteins often undergo a rapid degradation, not only within the body, but also during storage, if optimal conditions are not continuously maintained [13]. Furthermore, proteins are large molecules with a limited ability to cross biological membranes and reach the inner cell compartment [27, 33]. The oral route has obviously a great importance, since it is the most common way to deliver a drug to a patient, but in the case of protein-based drugs it involves several issues, such as the poor lipid solubility of many proteins, their low stability, and the rapid inactivation or enzymatic degradation in the gastrointestinal tract. Many strategies have been proposed to overcome these problems, like the addition of absorption enhancers, peptidase inhibitors, or of mucoadhesive delivery systems to the drug formulation. Enzymes used in drug formulations both for oral and intravenous administration very often need to be modified, to increase their stability and half-life, with the aim to cause an increase in their bioavailability, at the same time maintaining their therapeutic properties. One of the most common derivatization strategy is enzyme PEGylation (like in Oncaspar®, and Adagen®) [18, 48]. The PEGylation strategy can be very useful in case of oral administration, because the protein termini are protected from peptidase digestion, and the immunogenic response is reduced. On the other hand, this approach is not useful against some denaturing conditions, like low pH. Another strategy to improve protein stability is sequence modification, with the inclusion of unnatural amino acids (as for insulin by Eli Lilly or Proleukin by Chiron).

Hyperglycosilation is another common derivatization, as well as other conjugations with polysaccharide molecules [48]. An example of this kind of formulations is given by Ceredase®, a drug formulation of glucocerebrosidase. In this case, mannose moieties were attached to the enzyme. The mannose-terminated enzyme is efficiently delivered to the target cells, with a 100-fold increase of activity [5] with respect to the native one.

Another problem is the delivery to a specific tissue, like for all other drugs. The most promising approaches involve controlled-release formulations, in which the drug is entrapped within a scaffold structure, the carrier. In this way, the bioavailability of the molecule increases, because the half life depends on the carrier characteristics, and it can be released in the target tissue; furthermore, in many cases, the immunogenic response is avoided. Several types of carrier structures have been developed: among them, the most important ones are liposomes, polymeric micro- and nanoparticles, emulsions, and micelles. The mechanism of drug assimilation depends, in this case, on the composition of nanocarriers: the main uptake processes are mucoadhesion, cell internalization, or absorption from the gut lumen: For example, hydrophilic nanoparticles are mainly uptaken from the gut lymphoid tissue, while hydrophobic ones are mainly transported by enterocytes [21].

The most promising approaches are represented by emulsions, micro- and nanoparticles, and liposomes [35, 51, 54].

Microemulsions are composed by water, a lipophilic solvent, a surfactant and a co-surfactant, and are often used to enhance the oral bioavailability of drugs, including proteins and peptides [33]. Hydrophilic drugs can be easily internalized within microemulsions droplets, which also provide protection against proteolytic and chemical degradation in the intestinal lumen, when administrated orally. On the other hand, the synthesis of micro-and nanoemulsions presents also several disadvantages: it involves harsh conditions and denaturing agents, which can lead to processes like protein aggregation, denaturation, and oxidation at aqueous-organic solvent interfaces and cause a loss of activity for the encapsulated protein; to partially avoid these problems, some excipients or stabilizers can be added [33, 48].

Another important delivery tool is represented by microspheres, and, in particular, by microparticles constituted by poly-lactic and glycolic acid, joined by ester bonds (PLGA), which can protect the drugs from enzymatic or acidic degradation [10, 48, 54]. The size of these spheres can be tuned between 10 nm and 250 μm. Their biodegradation time can be varied with the lactic acid percentage, so that drug release takes place in weeks or months. In addition, they can be administrated both orally and via injection. Microspheres, like emulsions, are formed in harsh conditions, which include the presence of organic solvents, needed for the dissolution of polymers, and high-speed homogenization that can also cause protein denaturation. Another problem concerning PGLA microspheres is their degradation during the shelf life, which brings to the accumulation of lactic and glycolic acids in the storage solutions.

Several other polymeric micro-and nanoparticles have been developed for drug delivery, and in particular for protein delivery, such as polyglutamic acid (PGA), hybrid modified PEG polymers, and Medusa® polymers composed by glutamic acid and vitamin E, which have been applied to the delivery of insulin and growth hormone, among others.

Lipid drug delivery is another great promise for enzyme and protein therapies. In this field a large number of different carriers can be found, i.e. liposomes, solid lipid nanoparticles, micelles, lipid emulsions, microtubules, etc. Lipid-based vehicles exhibit a great versatility, and their characteristics can be tuned with the appropriate choice of the molecular characteristics of lipids [2, 33, 48].

The most important structures for protein delivery are liposomes and solid lipid nanoparticles. Liposomes present various advantages: they are biocompatible and they offer the possibility to deliver the drug directly within cells, or even inside cell compartments [55]. Liposomes are nowadays widely used in pharmaceutical formulations of great relevance, like Doxyl® for the delivery of the anticancer drug doxorubicin [29]. On the other hand, the clearance values of liposomes are strictly dependent on their size, and there is also the possibility of a rapid elimination from the blood. For instance, the delivery of enzymes entrapped within liposomes for ERT has been reported for beta-glucuronidase, showing a very rapid clearance from blood (4 min); on the other hand, the enzyme resulted active inside the lysosomes of liver and kidney cells after several days [53], suggesting a possible application for diseases involving these tissues, but not a specific delivery to other targets.

To conclude, all the most important protein delivery methods offer the possibility to enhance the drug bioavailability and the specific delivery, but also present important disadvantages, like the requirement for denaturing conditions during the synthesis, the use of organic solvents, and the need to use modified proteins, which implies a further synthetic step to obtain the final formulation, with a consequent increase in costs. In this section an alternative method will be presented, which does not present any of these issues. This method has also another peculiarity that makes it different from all the others, i.e. it has been designed to keep the drug active inside the scaffold, and not to deliver it to the tissues. The enzyme function is not inhibited by the presence of the vehicle, as will be illustrated in Chap. 7. In addition to protecting the enzyme from degradation, this approach avoids also its antigenicity.

5.3 Hydrogel Nanoparticles

An important class of nanoparticles is represented by those constituted by hydrogels.

Hydrogels were first introduced in the 1960s by Wichterle and Lim [61] and they have several pharmaceutical and biotechnology applications. Their unique ability to contain a large quantity of water within their structure, in addition to their low (or null) toxicity, tunable adhesion properties, and injectability makes them particularly suitable for biomedical and bioengineering applications.

The hydrogel structure is constituted by a polymeric cross-linked network, characterized by the presence of hyrophilic functional groups like hydroxyl or aminic moieties. Hydrogels for drug delivery systems can be constituted by natural polymers, like chitosan, hyaluronan, dextran or alginate, as well as synthetic ones, like polyacryamide, N-isoporopylacrylamide (NIPAAM), polyvinyl alcohol (PVA) or PEG [24, 40, 41, 44].

In the case of natural polymers, the biocompatibility can be higher, but the use of a synthetic polymer presents several advantages, i.e. the mechanical and structural properties (mesh size, molecular weight, viscosity, etc.) are more homogeneous and more easily tunable. Furthermore, synthetic polymers are in many cases biocompatible [24].

In most cases, hydrogels can be easily synthesized in relatively mild conditions, at room temperature and without the use of organic solvents [30], and this is particularly important to avoid protein denaturation.

The fundamental characteristic of hydrogels is that they are highly hydrophilic, and that they usually have the capability to swell in water; their physical and chemical characteristics, thus, are very similar to those of living tissues [4, 20, 45]. The swelling behavior is strictly related to the structural gel characteristics, and, in particular, is regulated by the degree of crosslinking, which determines the mesh size of the polymeric network. This parameter is therefore of fundamental importance in the release of a drug entrapped inside the gel structure.

Due to their sensitivity to external stimuli, hydrogels are excellent devices for controlled release: physically-induced release systems, for example, are sensitive to temperature, light, pressure, electric or magnetic fields, etc.; chemically-induced release gels, on the other hand, feature a response to changes of pH, solvent, ions, or to the presence of a specific molecule [6, 32, 49, 51].

All these characteristics make hydrogels extremely versatile tools. They are hydrophilic, water-absorbing devices with a tunable half-life, and with the possibility to be opportunely modified to interact with a specific target site. Furthermore, the nanoparticle accumulation is size-dependent [1, 31], and this property can be used to deliver the entrapped drug to a specific tissue. The biodistribution obviously depends also on the polymer composition, so different particles with the same size can be differently accumulated inside tissues, but generally small nanoparticles with 10–100 nm diameter have the longest blood permanence, while large nanoparticles, in the range of 500 nm, are accumulated in the liver.

5.4 Aim of the Work

A novel approach for protein-based drug formulations is presented here. The idea is to develop a nanometric cage formed by a hydrophilic polymeric network with a mesh size that blocks an enzyme, but allows diffusion of its substrates and products. In this way, the enzyme is entrapped in the nanoparticle (NP), and thus protected from degradation and antibodies, and at the same time can exert its catalytic activity. This approach requires the use of a biocompatible, but not biodegradable, polymer, and is different from the widely used methodology in which the enzyme is entrapped inside a device which is degraded inside the organisms, and exerts its activity only after it is released. Although the proposed principle is rather general, the enzyme chosen as a test case for this study is Cu, Zn superoxide dismutase (SOD) not only for its clinical relevance, that will be discussed in the next paragraph, but also for the small size of its substrates and products.

The second innovation of the proposed approach is aimed to solve the problem of protein denaturation under the conditions commonly employed to synthesize polymeric micro- or nano-particles. In order to synthesize the NPs under physiological conditions, liposomes will be exploited as nanotemplates. Phospholipid vesicles self-assemble spontaneously in water, and their dimension can be easily tuned by extrusion through filters with pores of appropriate size. The aqueous inner volume of liposomes represents a segregated environment, in which a chemical reaction can occur separately from the bulk solution. In our case, the reaction is the gel photopolymerization and leads to the formation of hydrogel nanoparticles (NPs) inside the lipid vesicles. Acrylamide (AA) was used as monomer for the realization of the hydrogel, with N, N′-methylene-bisacrylamide (MBA) as a crosslinker [3, 26]. Acrylamide-based hydrogels are widely used for drug delivery, thanks to their high biocompatibility and to the possibility to finely tune the mesh size, allowing the development of different types of applications [19, 41, 42, 46, 50].

The final product after the polymerization and purification processes is a device constituted by a hydrogel core, coated with a phospholipid bilayer.

5.5 Superoxide Dismutase and Disease

Superoxide dismutases (SOD) are a group of ubiquitous proteins, which catalyze the reaction of superoxide dismutation into water and hydrogen peroxide, according to the following equation [17, 34]:

$$2\,O_2\cdot^- + 2H^+ \rightarrow O_2 + H_2O_2$$

The catalytic center of the enzyme includes a metal ion, which takes part to the oxidation/reduction process.

The SOD family includes Cu, Zn SOD (used in this work), a cytosolic enzyme that is mostly present in eukaryotes, Fe-SOD, that is found in *E. coli* and other prokaryotes, Mn-SOD, a mitochondrial enzyme in eukaryotes, and Ni-SOD, that is found again in prokaryiotc organisms [47].

SOD is an enzyme of great biological relevance: superoxide is produced by the most important life processes, like aerobic metabolism, oxidative phosphorylation, mitochondrial electron flux, and photosynthesis. ROS are also generated by ionizing radiations. Unfortunately, the superoxide anion and the other oxygen reactive species, or ROS, are highly toxic for the organism: they are responsible for cell ageing and for a huge number of diseases. The ROS-mediated cellular damage includes ischaemic reperfusion injury, cancer, and DNA damage [37, 47]. Further studies have shown the relationship between oxidative stress and neurodegenerative disorders; a SOD mutation is often involved in the rise of amyotrophic lateral sclerosis (ALS) [37, 47] and the deficiency of mitochondrial SOD is associated with heart failure. In addition, deficiency of Cu, Zn SOD can lead to hepatocarcinogenesis and to an acceleration of skeletal muscle atrophy [16, 36]. Another important observation is that superoxide production increases in chronic or acute inflammation states. In this case, ROS can be secreted by activated neutrophils; if the endogenous SOD is not able to remove the excess of superoxide, ROS-mediated damage occurs. Thus, SOD has a great therapeutic potential for the treatment of neutrophil-mediated inflammation, also because it is involved in the mechanism of neutrophil apoptosis [62].

Preclinical studies on animal models revealed that SOD treatment has a protective effect against ischaemia-reperfusion injury, asthma, respiratory infections. Furthermore, SOD is involved in other regulation processes inside cells, and the increase of enzyme levels in all tissues could lead to metabolic alterations [25, 37]. Thus, it is very important to develop a tool to deliver the enzyme only to the diseased tissue. Human experimentation, with the use of the Orgotein formulation, i.e. bovine Cu,Zn SOD, has been carried out on inflammation-associated diseases like rheumatoid arthritis and osteoarthritis. For this purpose, SOD has also been encapsulated within liposomes [9]. However, after the clinical trials of the 1980s, the drug was retired from the market, mainly because of the immunological response against the enzyme, which was not human. To this end, various SOD-based formulations have been proposed, i.e. thermosensitive NiPAAM-HEMA hydrogels [28], biodegradable microspheres constituted by a PLGA-alginate-chitosan copolymer [22, 60], topical formulations [14], SOD-entrapping liposomes [7, 8]. Nevertheless, these systems have not been considered for clinical trials yet [23], and since they are all based on enzyme release they do not solve the immunogenicity problems. The same situation is observed for catalase, which converts the hydrogen peroxide produced by SOD to water and oxygen, concluding the process of superoxide degradation.

References

1. Alexis F, Pridgen E, Molnar LK, Farokhzad OC (2008) Factors affecting the clearance and biodistribution of polymeric nanoparticles. Mol Pharm 5:505–515
2. Almeida AJ, and Souto E (2007) Solid lipid nanoparticles as a drug delivery system for peptides and proteins. Adv Drug Deliv Rev 59:478–490
3. An SY, Bui MPN, Nam IJ, Kwi NH, Li LA, Choo J, Lee EY, Katoh S, Kumada Y, Seong GH (2009) Preparation of monodisperse and size-controlled poly(ethileneglycol) hydrogel nanoparticles using liposome templates. J Coll Interface Sci 331:98–103
4. Anseth KS, Bowman CN, Brannon-Peppas L (1996) Mechanical properties of hydrogel and their experimental determination. Biomaterials 17:1647–1657

5. Barranger JA, O'Rourke E (2001) Lesson learned from the development of enzyme therapy for Gaucher disease. J Inherit Metab Dis 24:89–96
6. Caykara T, Bulut M, Dilsiz N, Akyuz Y (2006) Macroporopus poly(acrylamide) Hydrogels: swelling and shrinking behaviors. Macromol Sci Part A Pure Appl Chem 43:889–897
7. Corvo ML, Boerman OC, Oyenm WJG, Van Bloois L, Cruz MEM, Crommelin DJA, Storm G (1999) Intravenous administration of superoxide dismutase entrapped in long-circulating liposomes II. In vivo fate in a rat model of adjuvant arthritis. Biochim Biophys Acta 1419:325–334
8. Corvo ML, Jorge JCS, van't Hof R, Cruz MEM, Crommelin DJA, Storm G (2002) Intravenous Superoxide dismutase entrapped in long-circulating liposomes: formulation design and therapeutic activity in rat adjuvant arthritis. Biochim Biophys Acta 1564:227–236
9. Cruz MEM, Gaspar MM, Martins MBF, Corvo ML (2005) Liposomal superoxide dismutases and their use in the treatment of experimental arthritis. Meth Enzymol 413:391–395
10. Dai C, Wang B, Zhao H (2005) Microencapsulation peptides and protein drugs delivery systems. Colloid Surf B 41:117–120
11. De Duve C (1966) The significance of lysosome in pathology and medicine. Proc Inst Med Chic 26:73–76
12. Desnick RJ, Schuchmann EH (2002) Enzyme replacement and enhancement therapies: lessons from lysosomial disorders. Nat Rev 3:954–966
13. Deutscher MP (2009) Maintaining protein stability. Meth Enzymol 463:121–127
14. Di Mambro V, Borin MF, Fonseca MJV (2003) Topical formulations with superoxide dismutase: influence of formulation composition on physical stability and enzymatic activity. J Pharm Biomed Anal 32:97–105
15. Edwards R (1999) Immunodiagnostics. Oxford University Press, Oxford
16. Elchuri S, Oberley TD, Qi W, Eisenstein RS, Roberts LJ, Van Remmen H, Epstein CJ, Huang TT (2005) Cu, Zn-SOD deficiency leads to persistent and widespread oxidative damage and hepatocarcinogenesis later in life. Oncogene 24:367–380
17. Falconi M, O'Neill P, Stroppolo ME, Desideri A (2002) Superoxide dismutase kinetics. Meth Enzymol 349:38–49
18. Frokjiaer S, Otzen DE (2004) Protein drug stability: a formulatioin challenge. Nature 4:298–306
19. Fuxman AM, McAuley KB, Schreiner LJ (2003) Modeling of free-radical crosslinking copolymerization of acrylamide and N, N'-methylenebisacrylamide for radiation dosimetry. Macromol Theory Simul 12:647–662
20. Ganji F, Vashegani-Farahani S, Vashegani-Farahani E (2010) Theoretical description of hydrogel swelling: a review. Iran Polym J 19:375–398
21. Garcia-Fuentes M, Prego C, Torres D, Alonso MJ (2005) A comparative study of the potential of solid triglyceride nanostructures coated with chitosan or poly (ethylene glycol) as carriers for oral calcitonin delivery. Eur J Pharm Sci 25:133–143
22. Giovagnoli S, Luca G, Casaburi I, Blasi P, Macchiarulo G, Ricci M, Calvitti M, Basta G, Calafiore R, Rossi C (2005) Long-term delivery of superoxide dismutase and catalase entrapped in poly(lactide-co-glycolide) microspheres: in vitro effects on isolated neonatal porcine pancreatic cell clusters. J Control Release 107:65–77
23. Grabowski GA, Hopkin RJ (2003) Enzyme therapy for lysosomal storage disease: principle, practice, and prospects. Annu Rev Genomics Hum Genet 4:403–436
24. Hamidi M, Azadi A, Rafiei P (2008) Hydrogel nanoparticles in drug delivery. Adv Drug Delivery Rev 60:1638–1649
25. Inoue M, Sato E, Nishikawa M, Park AM, Maeda K, Kasahara E (2002) Targeting superoxide dismutase to critical sites of action. Meth Enzymol 349:346–354
26. Kazakov S, Kaholek M, Teraoka I, Levon K (2002) UV-induced gelation on nanometer scale using liposome reactor. Macromolecules 35:1911–1920
27. Lee KY, Yuk SH (2007) Polymeric protein delivery systems. Prog Polym Sci 32:669–697
28. Li Z, Wang F, Roy S, Sen C, Guan J (2009) Injectable, highly flexible, and thermosensitive hydrogels capable of delivering superoxide dismutase. Biomacromolecules 10:3306–3316

29. Lian T, Ho RJY (2001) Trends and developments in liposome drug delivery systems. J Pharm Sci 90:667–680
30. Lin CC, Metters AT (2008) Hydrogels in controlled release formulations: network design and mathematical modeling. Adv Drug Delivery Rev 58:1379–1408
31. Liu D, Mori A, Huang L (1992) Role of liposome size and RES blockade in controlling biodistribution and tumor uptake of CM1-containing liposomes. Biochim Biophys Acta 1104:95–101
32. Makino K, Hiyoshi J, Ohshima H (2000) Kinetics of swelling and shrinking of poly (N-isopropylacrylamide) hydrogels at different temperatures. Colloid Surf B 19:197–204
33. Martins S, Sarmento B, Ferreira DC, Souto EB (2007) Lipid-based colloidal carriers for peptide and protein delivery-liposomes versus lipid nanoparticles. Int J Nanomed 2:596–607
34. McCord J (1988) Superoxide dismutase: the first twenty years (1968–1988). Free Radic Biol Med 5:363–369
35. Morishita M, Peppas NA (2006) Is the oral route possible for peptide and protein drug delivery? Drug Discov Today 11:905–910
36. Muller FL, Song W, Liu Y, Chaudhuri A, Sandra Pieke-Dahl S, Randy Strong RD, Huang TT, Epsteing CJ, Jackson Roberts II JR, Csete M, Faulkner JA, Van Remmen H (2006) Absence of CuZn superoxide dismutase leads to elevated oxidative stress and acceleration of age-dependent skeletal muscle atrophy. Free Radic Biol Med 40:1993–2004
37. Muscoli C, Cuzzocrea S, Riley DP, Zweier JL, Thiemermann C, Wang ZQ, Salvemini D (2003) On the selectivity of superoxide dismutase mimetics and its importance in pharmacological studies. Br J Pharmacol 140:445–460
38. Ngo T (2000) Biosensors and their application. Plenum Press, New York
39. O'Brady R, Schiffmann R (2004) Enzyme-replacement therapy for metabolic storage disorders. Lancet Neurol 3:752–756
40. Oh JK, Lee DI, Park JM (2009) Biopolymer-based microgels-nanogels for drug delivery applications. Progr Polym Sci 34:1261–1282
41. Oh JK, Drumright R, Siegwart DJ, Matujaszewski K (2008) The development of microgels/nanogels for drug delivery applications. Prog Polym Sci 33:448–477
42. Orakdogen N, Okay O (2006) Effect of initial monomer concentration on the equilibrium swelling and elasticity of hydrogels. Eur Polym J 42:955–960
43. Panke S, WUbbolts MG (2002) Enzyme technology and bioprocess engineering. Curr Opin Biotech 13:111–116
44. Patel HB, Patel HL, Shah ZH, Modasiya MK (2011) Review on hydrogel nanoparticles in drug delivery. Am J Pharm Tech Res 1:19–38
45. Peppas NA, Huang Y, Torres-Lugo M, Ward JH, Zhang J (2000) Physicochemical foundations and structural design of hydrogels in medicine and biology. Annu Rev Biomed Eng 2:9–29
46. Peppas NA (1997) Hydrogels and drug delivery. Curr Opin Colloid Interface Sci 2:531–537
47. Perry JJP, Shin DS, Getzoff ED, Tainer JA (2010) The structural biochemistry of superoxide dismutase. BBA 1804:245–262
48. Pisal DS, Kosloski MP, Balu-Iyer S (2010) Delivery of therapeutic proteins. J Pharm Sci 99:2557–2575
49. Qiu Y, Park K (2001) Environment-sensitive hydrogels for drug delivery. Adv Drug Delivery Rev 53:321–339
50. Sairam M, Babu VR, Naidu BVK, Aminabhavi TM (2006) Encapsulation efficiency and controlled release characteristics of crosslinked polyacrylamide nanoparticles. Int J Pharm 320:131–136
51. Schwall C, Banjeree IA (2009) Micro- and nanoscale hydrogel systems for drug delivery and tissue engineering. Materials 2:577–612
52. Siegel M, Bethune MT, Gass J, Ehren J, Xia J, Johannsen A, Stuge TB, Gray GM, Lee PP, Khosla C (2006) Rational design of combination enzyme therapy for celiac sprue. Chem Biol 13:649–658

53. Steger L, Desnick RJ (1977) Enzyme therapy VI: comparative in vivo fates and effect on lysosomial integrity of enzyme entrapped in negatively and positively charged liposomes. Biochim Biophys Acta 464:530–546
54. Teply BA, Tong R, Jeong SY, Luther G, Sherifi I, Yim CH, Khademhosseini A, Farokhzad OC, Langer R, Cheng J (2008) The use of charge-coupled polymeric microparticles and micromagnets for modulating the bioavailability of orally delivered macromolecules. Biomaterials 29:1216–1223
55. Torchilin VP (2005) Recent advantages with liposomes as pharmaceutical carriers. Nat Rev Drug Discov 4:145–160
56. Van Capelle CI, Winke LPF, Hagemans MLC, Shapira SK, Arts WFM, Van Doorn PA, Hop WCJ, Reuser AJJ, van der Ploeg AT (2008) Eight years experience with enzyme replacement therapy in two children and one adult with Pompe disease. Neuromuscul Disord 18:447–452
57. Vellard M (2003) The enzyme as drug: application of enzymes as pharmaceuticals. Curr Opin Biotechnol 14:444–450
58. Walsh G (2000) Biopharmaceutical benchmarks. Nat Biotechnol 18:831–833
59. Walsh G (2006) Directory of therapeutic enzymes. CRC Press, London
60. Wang Y, Gao JQ, Zheng CH, Xu DH, Liang WQ (2006) Biodegradable and complexed microspheres used for sustained delivery and activity protection of SOD. J Biomed Mater Res Part B Appl Biomater 79B:74–78
61. Wichterle O, Lim D (1960) Hydrophilic gels for biological use. Nature 185:117–118
62. Yasui K, Baba A (2006) Therapeutic potential of superoxide dismutase (SOD) for resolution of imflammation. Inflamm Res 55:359–363

Chapter 6
Materials and Methods

6.1 Materials

AA, MBA and sodium dodecyl sulphate (SDS) were purchased from Fluka. Di-etoxyacetophenone (DEAP), ovalbumin, and pyrogallol were purchased from Sigma-Aldrich.

Egg phosphatidyl choline was purchased from Sigma-Aldrich. All the other phospholipids were purchased from Avanti Polar Lipids.

Bovine Cu, Zn-superoxide dismutase was a kind gift by Prof. Giorgio Ricci.

6.2 Nanoparticles Synthesis

Egg-PC lipids were dissolved in a chloroform:methanol 1:1 mixture, in a lipid amount such to obtain a such to obtain a 100 mM concentration in the final liposome stock solution. The solvent was then gently evaporated in a rotary vacuum system, until a homogeneous, thin film was formed. Complete solvent evaporation was ensured by applying a rotary vacuum pump for at least 2 h. The lipid film was then hydrated with a 10 mM phosphate buffer (pH 7.4) solution containing AA (190.1 mg/ml), MBA (10.4 mg/ml) (DEAP, 0.0022 mg/ml), and eventually SOD (4 mg/ml). After 10 freeze/thaw cycles liposomes were extruded through two stacked polycarbonate membranes with a nominal pore diameter of 400 or 200 nm, using a LiposoFast extruder (Avestin, Mannheim, Germany). After extrusion, the sample was diluted 20-fold with a glucose solution of the same osmolarity of the hydration solution. Since the inner aqueous volume of the vesicles is unperturbed by this dilution, this procedure avoids polymerization in the outer solution only [6]. The sample was then irradiated for 10 min in a photoreactor equipped with 10 lamps at 367 nm, for a total power of 1,500 W. When the enzyme was included into the formulation, the irradiation process was carried out placing the sample on an ice bath, to limit the temperature increase and thus avoid enzyme denaturation.

S. Bobone, *Peptide and Protein Interaction with Membrane Systems*, Springer Theses, DOI: 10.1007/978-3-319-06434-5_6,

After irradiation, the sample was purified by dialysis using 300,000 MWCO SpectraPor® membranes (Spectrum Europe, The Netherlands), which are permeable to the unentrapped protein, whose MW is 32,000. Before the purification process, the lipid membrane was removed by the addition of the detergent SDS at a detergent to lipid ratio of 5 to 1 [5, 8]. For SOD containing NPs, dialysis was carried out at 4 °C.

6.3 Light Scattering Measurements

To characterize the NPs size, light scattering measurements were performed using a Horiba LA 300 instrument, equipped with a laser diode of $\lambda_{em} = 650$ nm, of 5 mW potency. Experiments were performed before and after membrane solubilization.

6.4 AFM Measurements

For AFM experiments, an amount of NPs stock solution was deposed on mica supports into a clean room and dried overnight. The same process was carried out using 1:10 and 1:100 diluted solutions. Measurements were carried out using a Nanoscope IIIA microscope (Bruker), in tapping mode, using Silicon PPP-NCH tips (height 10–15 μm, Kel 42 N/M, curvature radius $\leq$7 nm).

6.5 Enzymatic Activity Assay

SOD activity was determined using the pyrogallol assay [3, 7].

Pyrogallol is a molecule that rapidly auto-oxidizes in an alkaline environment, becoming yellow. The auto-oxidation rate can be measured by following the increase in absorbance at 420 nm. The oxidation process, is inhibited by the presence of SOD; therefore, this reaction can be exploited to measure SOD activity.

For the activity assay, a TRIS-HCl buffer (20 mM) at pH 8.2 was used. The blank solution was constituted by 990 μl of buffer and 10 μl of a 20 mM pyrogallol solution (pH = 3). The absorbance variation of this sample (solution A) was followed for 3 min immediately after pyrogallol addition.

The sample of interest (solution B) was prepared adding to the TRIS buffer an appropriate amount of NPs-containing solution, to reach a final volume of 990 μl. Then, 10 μl of the pyrogallol solution were added and the absorbance variation was followed as for solution A. An example of absorbance variation obtained with and without SOD is shown in Fig. 6.1.

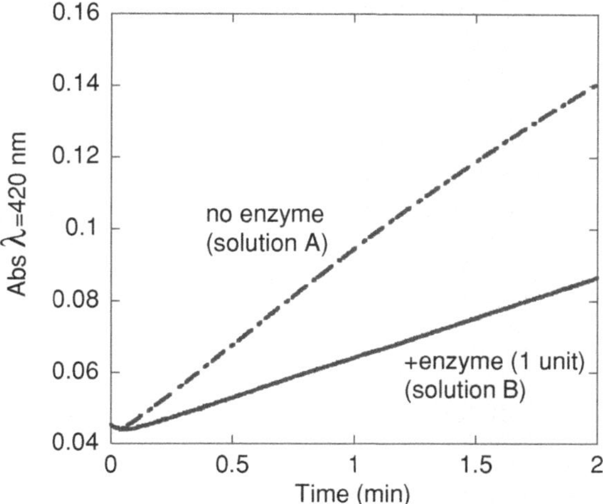

Fig. 6.1 Pyrogallol auto-oxidation kinetics in the absence (*continuous line*) and in the presence (*dotted line*) of one unit of SOD

In both cases the curve slope was calculated between 60 and 120 s (S_A and S_B). An enzyme unit is defined as the quantity of protein inhibiting the reaction down to 50 % of its spontaneous rate ($S_B = S_A/2$). The condition $0.35\ S_B < S_A < 0.65\ S_B$ must be verified to obtain a value in the linear range, to extrapolate the number of enzyme units in the solution B. Under these conditions, SOD units are calculated from the ratio $\frac{\frac{SA}{2}}{SB}$.

6.6 Protein Labeling

Ovalbumin was incubated with FITC at pH 9.3 for 12 h and then separated from the free dye on a Sephadex-G25 column. The labeling ratio was calculated from the UV-visible spectrum maxima at 490 and 280 nm according to the following equation

$$A_{eff}(280\,\mathrm{nm}) = \frac{A\exp(280) - A\exp(495)}{rFITC}$$

where $rFITC = 3$ [2] and considering that for ovalbumin $\varepsilon_{280\ \mathrm{nm}} = 32{,}000\ \mathrm{M}^{-1}\ \mathrm{cm}^{-1}$. FITC concentration in the sample was calculated from its absorption at 495 nm ($\varepsilon_{495\ \mathrm{nm}} = 63{,}000\ \mathrm{M}^{-1}\ \mathrm{cm}^{-1}$).

6.7 Fluorescence Measurements

Fluorescence spectra were performed with a SPEX Fluoromax-2 spectrofluorometer. SOD spectra were collected using $\lambda_{exc} = 270$ nm in a spectral range between 290 and 400 nm, using a cutoff emission filter at 295 nm.

Spectra of Rho-labeled lipid vesicles were collected using $\lambda_{exc} = 560$ nm in a spectral range between 570 and 640 nm in a 10 × 10 mm cell (bandwidth: 1 nm excitation, 3 nm emission).

Fluorescence anisotropy measurements of SOD were carried out, at $\lambda_{exc} = 270$ nm and $\lambda_{em} = 322$ nm in a 10 × 10 mm cell (bandwidth: 15 nm excitation, 20 nm emission).

FRET excitation spectra were collected between 270 and 400 nm using $\lambda_{em} = 480$ nm, in a 5 × 10 mm cell (bandwidth: 5 nm excitation, 5 nm emission).

6.8 Fluorescence Recovery After Photobleaching

Fluorescence recovery after photobleaching (FRAP) experiments were carried out with an Olympus confocal microscope (FluoView 1000) on a bulk gel of the same composition than NPs. Measurements were performed using a 20x objective. The region of interest had a diameter of 40 μm.

40 snapshots were collected before photobleaching to operate a correction for the excitation laser-induced bleaching, considering the mean decrease of fluorescence in three different regions of interest of the same size. At this point, the photobleaching was performed and after a 2 s delay 450 snapshots (320 × 320 pixels in dimensions) were collected, for a time of 900 s. The scan speed was 8 us/ pixel, The time of scan was of 2 s. The photobleaching time was of 2,000 ms.

6.9 Enzyme Protection from Thermal Degradation

A SOD-containing NPs sample and a SOD solution were heated at T = 80 °C into an AccuBlock digital dry bath (Labnet International). The activity of the two samples was measured at different times.

6.10 Synthesis of FLAC

Fluorescent NPs have been synthesized, to allow their direct visualization in cell-uptake experiments. For this purpose, the fluorescent analogue of acrylamide, fluorescein acrylamide (FLAC) (Fig. 6.3) was synthesized, starting from acryloyl chloride and fluorescein amine (Fig. 6.2) [9].

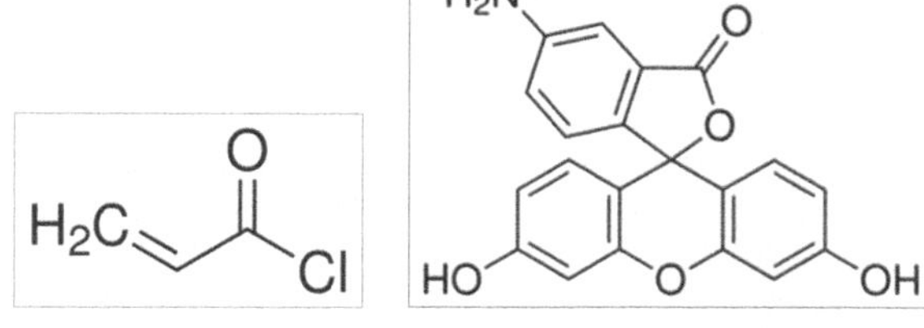

Fig. 6.2 Structures of acryloyl chloride (*left*) and fluorescein amine (*right*)

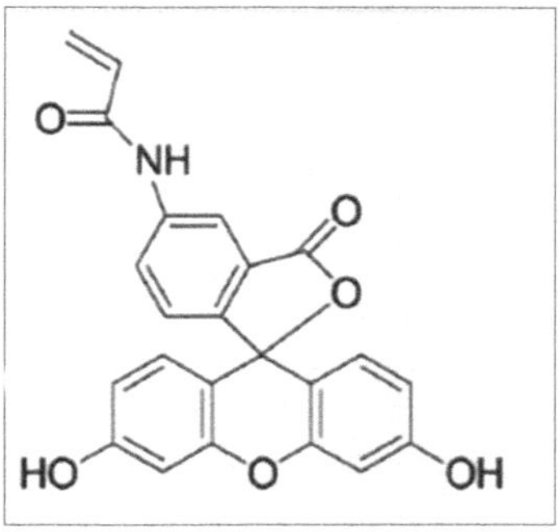

Fig. 6.3 Structure of FLAC

Fluorescein amine was dissolved in anhydrous acetone in a 3-neck flask; acryloyl chloride was then added. The reaction was carried out under an Ar stream, for 3 h, until the precipitation of product was complete. The FLAC powder was then dried under vacuum and used without any further purification. The product was obtained with a yield of 70 %.

6.11 Nanoparticles Uptake Experiments

To allow the visualization of NPs, fluorescein- labeled NPs were synthesized, using 5 % of FLAC. Murine fibroblasts (NIH 3T3 line) were grown at 37 °C in a 5 % CO_2 atmosphere in the Dulbecco's modified Eagle Medium (DMEM), with the addition of fetal bovine serum (FBS) 10 %, glutamine 20 μM and 100 U/ml of penicillin.

The NPs solution was sterilized at 120 °C in autoclave for 30 min and then incubated with cells for 72 h. Cells were treated with the dye FAST DiI (Fig. 6.4) (Invitrogen), which binds to the lipid membranes and allow their visualization ($\lambda_{exc} = 549$ nm; $\lambda_{em} = 565$ nm) images were then acquired with a Nikon Eclipse TE 2000-S confocal microscope. The same microscope was used for all the following experiments.

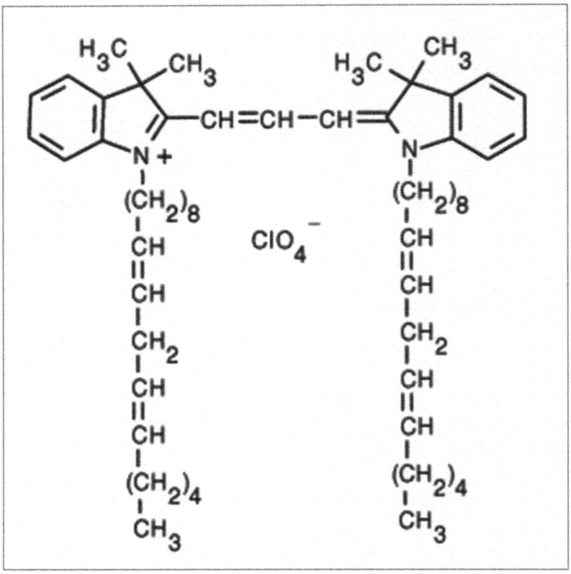

Fig. 6.4 DiI structure

6.12 Nanoparticles Biocompatibility Experiments

For cell viability assays, the NPs concentration has been estimated from the lipid concentration, calculating the concentration of spherical vesicles entrapping the NPs from the following equation:

$$[\text{liposomes}] = [\text{L}]/\text{R}$$

where [L] is the lipid concentration and R is given by the ratio $\frac{2A_{\text{vesicles}}}{A_{\text{lipid}}}$.

A_{vesicles} is the surface area of a liposome (calculated from its radius) and A_{lipid} is the area a lipid molecule occupies in a bilayer (assumed equal to 70 Å^2) [10].

6.12.1 Trypan Blue Assay

The Trypan blue (TB) assay reveals the integrity of cell membranes, and thus, it is a measurement of cells viability [1]. The chromophore is negatively charged (Fig. 6.5) and is not able to cross the membrane lipid bilayer of a healthy cell. If the membrane is damaged, instead, the dye penetrates inside the cell, which appears blue.

Cell viability was checked after 72 h of incubation with NPs. A 0.4 % TB solution (0.1 ml) was added to cells; samples were stirred for 15 min and then live and dead cells were visualized with an optical microscope and counted with the help of a Neubauer chamber. Results were then compared with those obtained for a control culture of untreated cells.

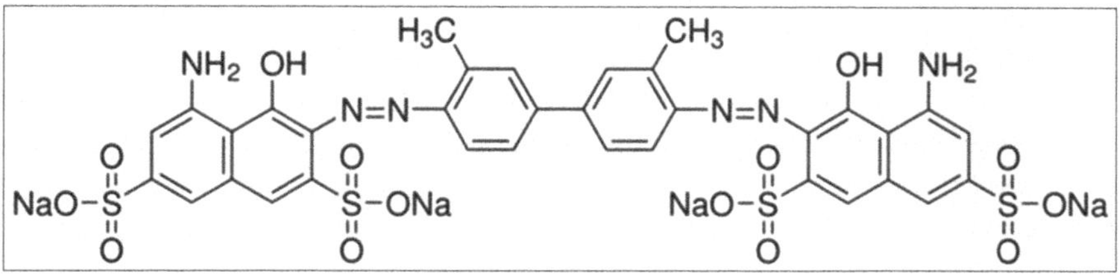

Fig. 6.5 Trypan blue structure

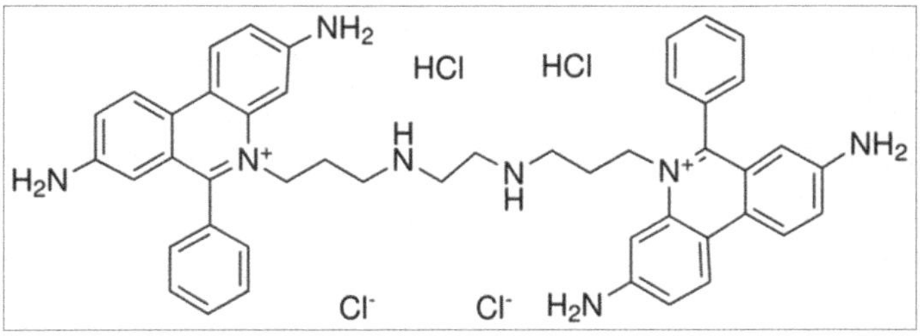

Fig. 6.6 EthD-1 homodimer

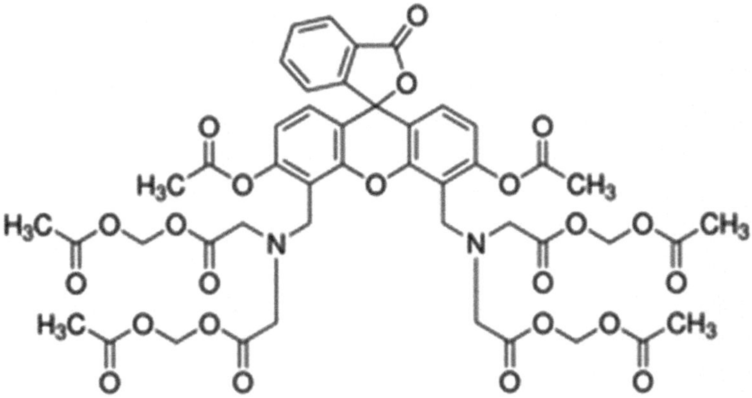

Fig. 6.7 Calcein AM structure

6.12.2 Live or Dead Assay

For the live or dead experiment, fibroblasts were incubated with NPs (0.4 nM). Cells were then treated with two fluorophores, the calcein acteoxy-metyl (AM) and the ethidium homodimer-1 (EthD-1, Fig. 6.6). Calcein AM (Fig. 6.7) is a cell-penetrating molecule, which is converted in the highly fluorescent calcein by the ubiquitous intracellular enzyme esterase. In this case, cells appear green (the excitation and emission wavelength of the dye are 495 and 515 nm). On the other hand, EthD-1 enters only cells with damaged membranes and undergoes a 40-fold fluorescence enhancement upon binding with nucleic acids. In this case, dead cells appear red [4]. Cells treated with NPs, as well as a control culture, were visualized after 24 h.

References

1. Altman SA, Randers L, Rao G (1993) Comparison of trypan blue dye exclusion and fluorometric assays for mammalian cell viability determinations. Biotechnol Prog 9:671–674
2. DePetris S (1978) Methods in membrane biology, vol 9. Plenum Press, New York, pp 1–201
3. Falconi M, O'Neill P, Stroppolo ME, Desideri A (2002) Superoxide dismutase kinetics. Meth Enzymol 349:38–49
4. Haugland RP, MacCoubrey IC, Moore PL (1994) US Patent 5,314,805
5. Helenius A, Simons K (1975) Solubilization of membranes by detergent. Biochim Biophys Acta 415:29–79
6. Kazakov S, Kaholek M, Teraoka I, Levon K (2002) UV-induced gelation on nanometer scale using liposome reactor. Macromolecules 35:1911–1920
7. Marklund S, Marklund G (1974) Involvement of the superoxide anion radical in the autoxidation of pyrogallol and a convenient assay for superoxide dismutase. Eur J Biochem 47:469–474
8. Pierce Biotechnology Inc. (2007) Remove detergent from protein samples. Tech Tip 19
9. Sun H, Scharff-Poulsen AM, Gu H, Almdal K (2006) Synthesis and characterization of radiometric, pH-sensimg nanoparticles with covalently attached fluorescent dyes. Chem Mater 18:3381–3384
10. Torchilin V, Weissing V (eds) (2003) Liposomes: a practical approach. Oxford University Press, Oxford

Chapter 7
Results and Discussion

7.1 Hydrogel Formulation

Acrylamide (AA) was chosen as the building block of the polymer used in the preparation of NPs. Polyacrylamide gels have been extensively characterized in the past years [4, 5, 18]. The crosslinking molecule, used to create the network, was N,N′-methylene-bisacrylamide (MBA). This monomer/crosslinker system is widely used for biomedical applications, because the mesh size of the polymeric network can be easily tuned; in addition, the photopolymerization is a fast and easy reaction, which only needs the addition of a very small amount of photoinitiator, DEAP in our case. The structures of monomer, crosslinker and photoinitiator are reported in Fig. 7.1.

The composition of polyacrylamide hydrogels is often defined by employing two parameters, T and C. T represents the monomer concentration, while C is the weight fraction of crosslinker, defined according to the following equations:

$$\mathrm{T} = \frac{monomer\, weight\, (g)}{solvent\, volume\, (ml)} \times 100$$

$$\mathrm{C} = \frac{crosslinker\, weight\, (g)}{monomer + crosslinker\, weight\, (g)} \times 100$$

Previous studies on acrylamide-based hydrogels [2] showed that a fixed monomer concentration (T), the gel pore size is directly dependent on the amount of crosslinker (C), but also that a threshold exists, corresponding to approximately C = 5 %. Above this value, bisacrylamide starts to aggregate, leading to the formation of inhomogeneities. On the other hand, pore size is an inverse function of T, but this dependence becomes weaker above T = 20 %, substantially reaching a plateau.

Preliminary experiments were carried out using a T = 5 %, C = 3 % formulation, but in this case the hydrogel mesh size was too wide to efficiently entrap the SOD molecules, which diffused very rapidly out of the nanoparticles (data not shown). To

S. Bobone, *Peptide and Protein Interaction with Membrane Systems*,
Springer Theses, DOI: 10.1007/978-3-319-06434-5_7,

Fig. 7.1 Structures of AA (*top*, *left*), MBA (*top*, *right*), and DEAP (*bottom*)

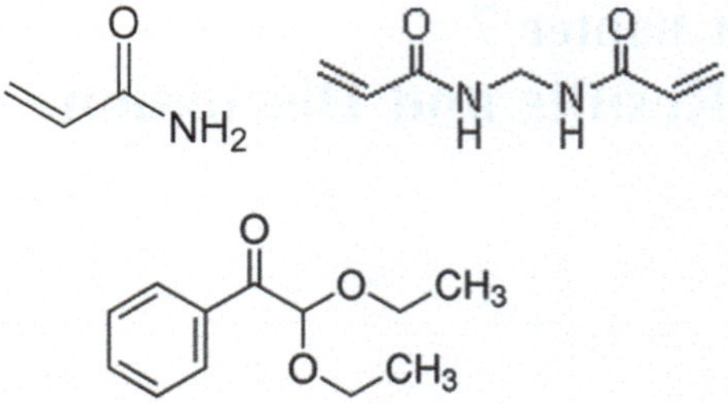

Table 7.1 NPs composition at T = 20 %, C = 5 % and T = 5 %, C = 3 % formulation

Molecule	Concentration (mM) T=20%, C=5%	Concentration (mM) T=5%, C=3%
ePC	100	10
AA	2674	660
MBA	67.4	33
DEAP	22	22
SOD	0.256	0.512
Total concentration of polymerization solution	2785	725

solve this problem, the hydrogel formulation was optimized with the following parameters: T = 20 %, C = 5 %, which were previously shown to minimize the mesh size [16, 21] and thus represent the best condition to entrap the enzyme inside the NPs. The monomer and crosslinker amounts are summarized in Table 7.1.

7.2 Bulk Polymerization

Before the NPs synthesis, a bulk gel was prepared to check the effective gelation of the solution: the presence of the radical scavenger SOD could inhibit the polymerization, but this did not occur, at least at the investigated concentration (Fig. 7.2).

7.2.1 SOD-Membrane Interaction: Fluorescence Anisotropy

As shown in the first part of this thesis, several peptides and proteins exhibit a strong membrane-perturbing activity. Therefore, in principle, a strong interaction between SOD and the lipid bilayer could perturb the membrane permeability, thus

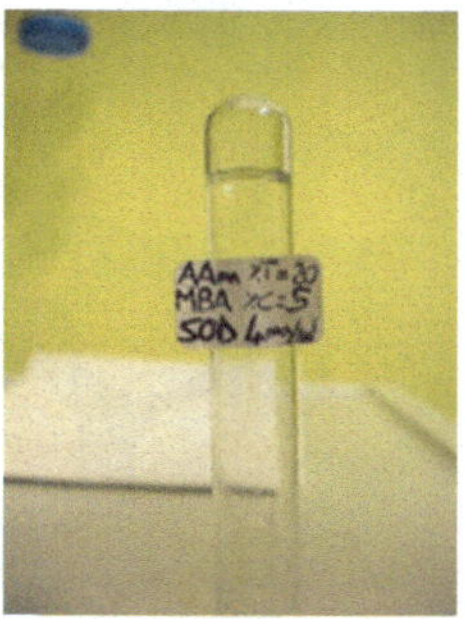

Fig. 7.2 Hydrogel bulk polymerization in the presence of SOD

Table 7.2 Anisotropy measurements of SOD. Experimental conditions: $\lambda_{exc} = 270$ nm; $\lambda_{em} = 305$ nm; slits: cutoff filter = 295 nm; cell size: 5 mm excitation side, 10 mm emission side

Sample	Anisotropy
Free SOD (10 μM)	0.10 ±0.01
SOD + liposomes	0.11 ±0.01

frustrating our approach. To rule out this hypothesis, fluorescence anisotropy measurements were performed, exploiting the intrinsic fluorescence of bovine SOD, due to a Tyr residue in its sequence. A first measurement was carried out on a 10 μM SOD solution; then, a volume of an ePC liposome solution was added to the sample, to reach a final lipid concentration of 100 μM, thus, in a large excess. If the protein binds to the lipid vesicles, an anisotropy variation should be observed. By contrast, SOD anisotropy was unaffected by the liposome addition, as reported in Table 7.2. On the basis of this experiment, a significant interaction between protein and membrane can be excluded.

7.2.2 SOD-Membrane Interaction: FRET Measurements

In order to definitely exclude a membrane-protein interaction, a FRET experiment was performed. In this case, liposomes were prepared including the DPH probe (0.5 %) in the lipid bilayer. No FRET was detected between the Tyr residue of the enzyme (the donor) and the DPH probe (the acceptor), proving that SOD does not interact with the lipid bilayer (data not shown).

7.3 Synthesis Protocol

The main steps of the synthesis protocol are represented in Fig. 7.3.

Step 1: Lipid film preparation and hydration. Egg PC is evaporated in a rotatory vacuum system, until a thin film is formed on the wall of the round bottom flask.

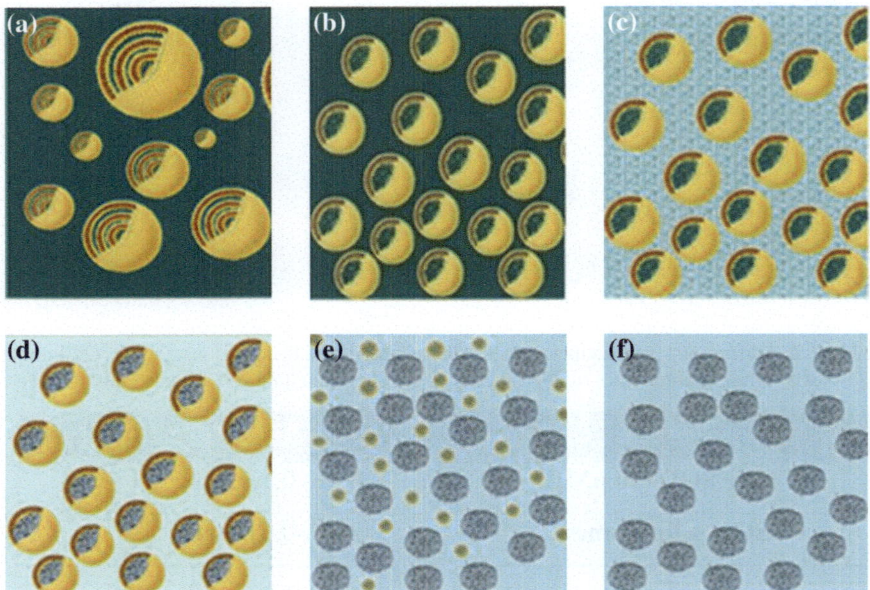

Fig. 7.3 Summary of NPs synthesis in different steps. **a** Lipid film hydration with the polymerization solution and formation of multilamellar vesicles. **b** Liposomes extrusion. **c** Dilution of the external solution. **d** Polymerization of the inner solution. **e** Lipid membrane removal. **f** NPs purification

The film is then hydrated with the mixture of monomer, crosslinker, photoinitiator and, if needed, SOD. This step leads to the formation of multilamellar vesicles. To facilitate the formation of unilamellar vesicles, 10 freeze and thaw cycles are performed.

Step 2: Extrusion. The size of liposomes is selected by extruding the solution through two polycarbonate membranes with well-defined pore size (which can be selected between 50 and 400 nm).

Step 3: Dilution. The vesicles suspension is diluted 20 times with a glucose solution of the same osmolarity of the hydration solution (2.8 M). The dilution avoids the bulk polymerization, and allows the gel formation only inside the lipid vesicles.

Step 4: Photoirradiation. The solution is irradiated with UV light to allow the photopolymerization of the solution entrapped inside liposomes. In this step, vesicles act as nanoreactors for the polymerization. In this step, membrane-coated NPs are obtained.

Step 5: Membrane removal. The lipid membrane can be removed by adding a detergent, sodium dodecyl sulfate, to the NPs suspension. The NPs obtained in this step will be called "naked".

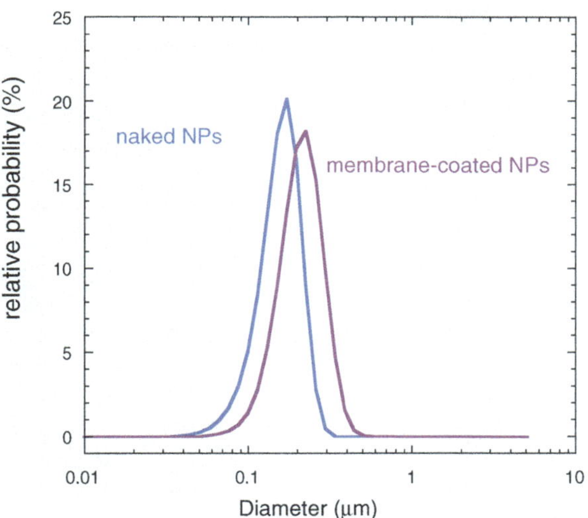

Fig. 7.4 Light scattering profiles of a NPs sample before and after the lipid membrane removal

Step 6: Sample purification. Lipid molecules, as well as unreacted monomers and crosslinker, and unentrapped enzymes are removed by dialysis.

It is worth to underline that the entire process occurs under physiological condition, without the use of any organic solvent or other harsh conditions that can reduce the enzymatic activity.

7.4 Nanoparticles Characterization

7.4.1 Light Scattering Characterization

The NPs size can be tuned by extrusion, and was characterized by light scattering experiments before (coated NPs) and after (naked NPs) membrane removal. In this work, batches of NPs have been prepared with a nominal diameter of 200 or 400 nm; their real size was determined by light scattering measurements (Fig. 7.4). For all preparations the size distribution was rather narrow, with a typical standard deviation around 40 nm.

The size of naked NPs was slightly smaller than that of the coated ones. The lack of swelling [7, 9] is to be expected for a highly crosslinked hydrogel [4, 10, 13]. The slight decrease in size caused by membrane removal (which however is definitely larger than the 4 nm thickness of a lipid bilayer), suggests that the polymerization leads to the formation of particles which do not occupy the entire inner volume of liposomes.

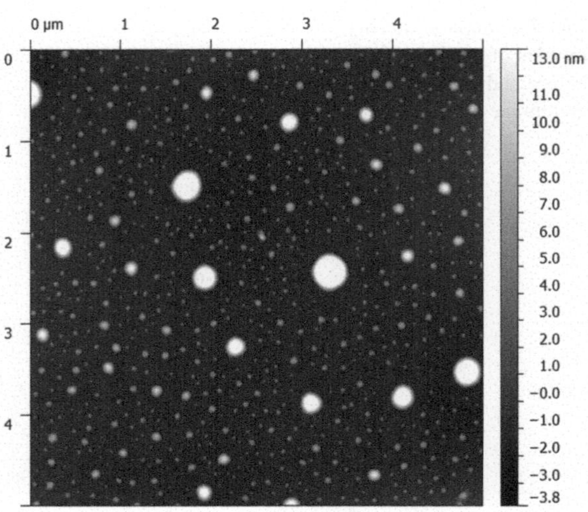

Fig. 7.5 AFM image of an uncoated NPs sample prepared from a 100-fold diluted solution

7.4.2 AFM Characterization

AFM experiments were also carried out, allowing a direct visualization of NPs. Images were collected from different samples, obtained from NPs of diluted solutions (100 times from the stock, for the best conditions) to avoid aggregation effects, in several different regions of the mica support. Experiments were performed on both membrane-coated and naked NPs.

AFM images showed a large number of well-separated nanoparticles on every explored region. Two representative images are reported in Figs. 7.5 and 7.6.

Images acquired from samples constituted of membrane-coated NPs showed that the membrane lost its integrity during the deposition and drying process and that the lipid molecules tended to attach to the surface around the particles (Fig. 7.6); the analysis of the zones formed by lipids shows very reproducible height profiles (Fig. 7.7), suggesting that these regions represent a lipid bilayer.

The size of the NPs sample used for AFM experiment has been characterized by light scattering measurements. The distribution profiles before and after the membrane solubilization are reported in Fig. 7.8. However, size distribution profiles have been calculated also from AFM images (in Fig. 7.9 the results obtained for naked NPs are reported). In this case, results are referred to dried NPs, which became smaller.

7.5 Membrane Removal by Detergent Addition

The removal of the lipid membrane from NPs was obtained by the addition of sodium dodecyl sulfate (SDS) in a lipid to detergent ratio of 5:1. A first proof of the effective membrane solubilization can be obtained from the light scattering

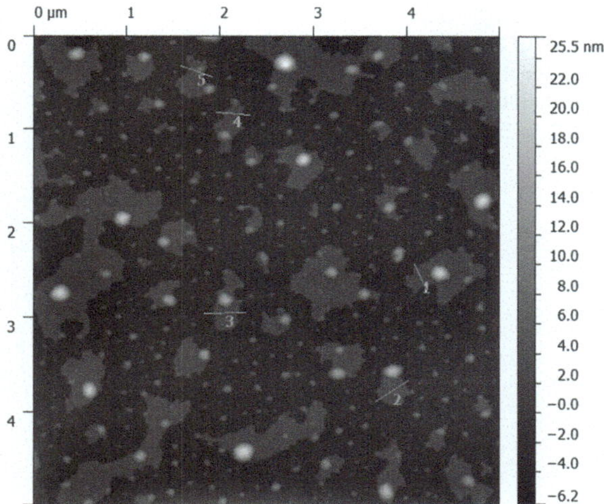

Fig. 7.6 AFM image of a NPs sample prepared from a 100-fold diluted solution from the stock. The NPs in the sample were analyzed with the intact lipid membrane. The *lines* in the pictures represent the regions whose profiles have been analyzed in Fig. 7.7

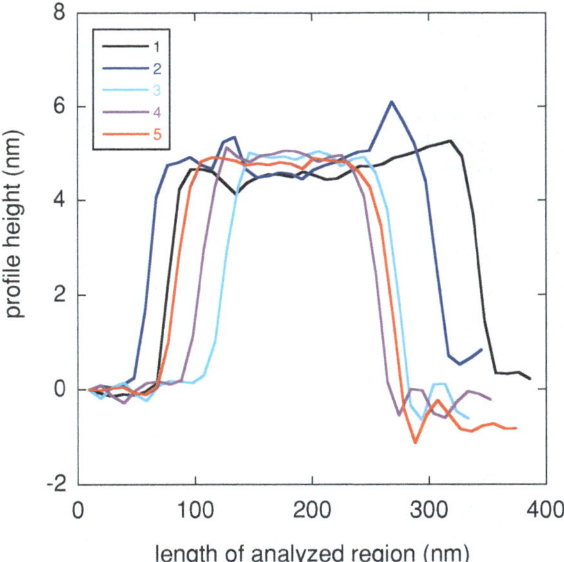

Fig. 7.7 Height profiles of the regions indicated in this figure. The height profiles of the zones are reported as a function of the distance

measurements showing a reduction of NPs size, but a specific experiment was also performed. NPs were prepared including in the lipid film a small amount of fluorescent lipid Rho-PE (1 % of total lipids).

The fluorescence signal was recorded immediately after the preparation; the sample was then treated with detergent and then submitted to an extensive dialysis cycle, to remove the solubilized lipids. After the purification, a fluorescence spectrum was collected again, showing a significant signal reduction, that is a clear evidence of lipid membrane removal (Fig. 7.10).

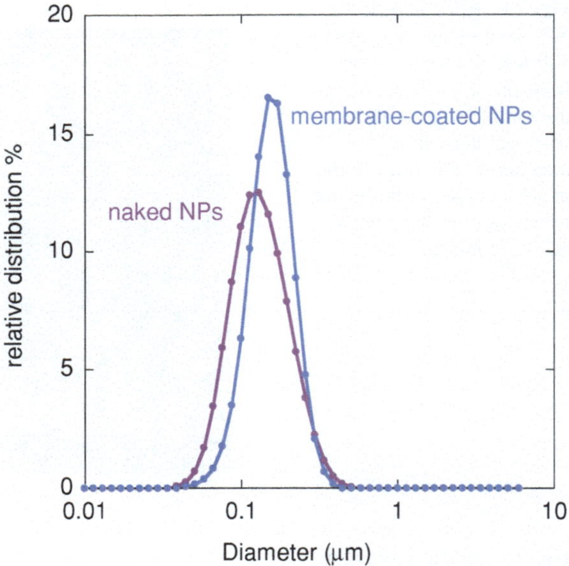

Fig. 7.8 Light scattering profiles of NPs used for AFM experiments. NPs gel formulation: T = 20 %, C = 5 %

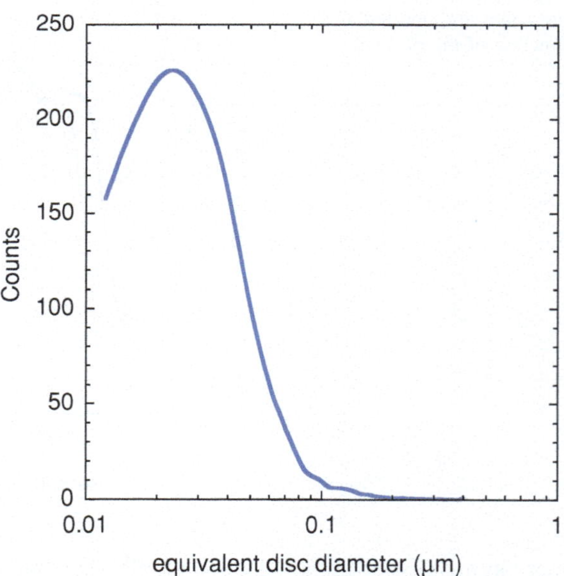

Fig. 7.9 Size distribution profiles obtained for naked NPs from AFM measurements

7.6 Protein Labeling for FRAP Experiments

In order to assess the ability of the gel network to efficiently block the SOD molecules, fluorescence recovery after photobleaching (FRAP) experiments were performed. The FRAP technique is a powerful tool to determine the degree of

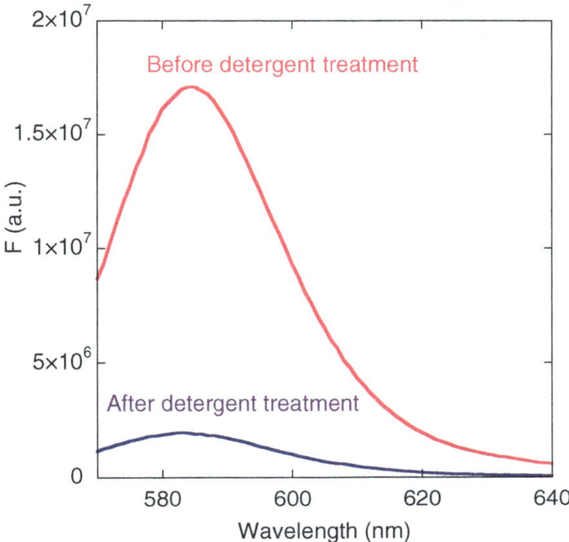

Fig. 7.10 Fluorescence spectra of Rho-labeled NPs before and after detergent addition. Experimental conditions: $\lambda_{exc} = 560$ nm; bandwidth:1 nm in excitation, 3 nm in emission; cell size 1 × 1 cm

mobility of macromolecules and their diffusion coefficients [3, 8, 25]. In this technique, a region of a fluorescent sample is bleached by an intense light pulse, and the recovery of the signal, due to the diffusion of other fluorescent molecules into the bleached region, is recorded. This allows the determination of the diffusion coefficient of the fluorescent species. The possible presence of immobile species is detected by the lack of complete recovery of the fluorescence signal.

FRAP experiments were carried out on a model protein, FITC-labeled ovalbumin. The dimensions of this protein are comparable to those of SOD: its molecular weight is 42,700 KDa (the MW of the SOD dimer is 32,500), and its hydrodynamic radius is 30.5 Å [24], as compared to 28.5 Å for SOD [22], but the cost of ovalbumin is considerably lower.

The protein was labeled according to a standard protocol, described in Sect. 6.6. The protein concentration after purification was calculated from the absorption at 280 nm (Fig. 7.11), as described in Sect. 6.6.

7.7 Fluorescence Recovery After Photobleaching

To perform FRAP experiments, a bulk polyacrylamide gel was prepared, of the same composition used for NPs preparation (T = 20 %, C = 5 %), in which FITC-ovalbumin was entrapped, at 30 μM concentration. After photopolymerization, the hydrogel was left in water and extensively rinsed until the washing solution did not show any fluorescence. FRAP measurements were then performed, by cutting a thin gel slice of about 2 mm and positioning it on the microscope observation glass slide. The recovery profile obtained in these conditions is reported in Fig. 7.12.

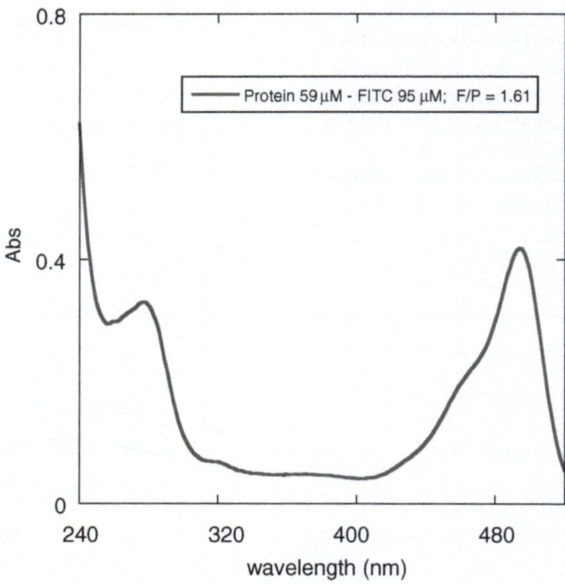

Fig. 7.11 UV-Vis spectrum of FITC-labeled ovalbumin. Pathlength = 0.1 cm

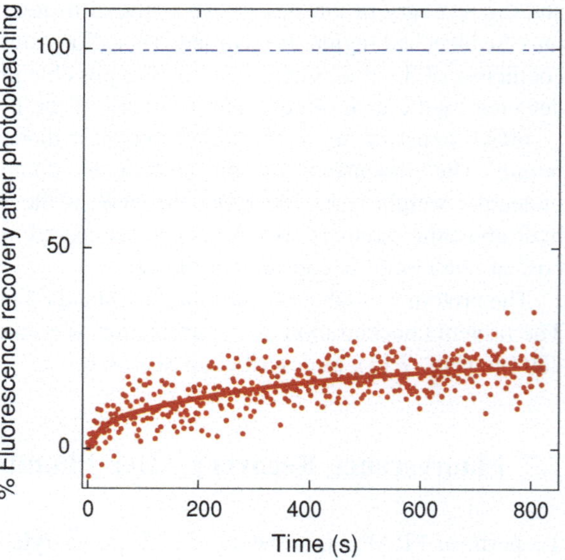

Fig. 7.12 FRAP of protein entrapped inside the T = 20 %, C = 5 % polyacrylamide hydrogel. Experimental conditions: λ_{exc} = 490 nm; Region of interest (ROI) diameter = 40 μm

As shown in Fig. 7.12, a major fraction of the protein (about 80 %) was not able to participate to the fluorescence recovery, indicating that it was efficiently blocked within the gel network, at least in the time range investigated. On the other hand, a small mobile fraction was present, and it is responsible for a recovery of about 20 %.

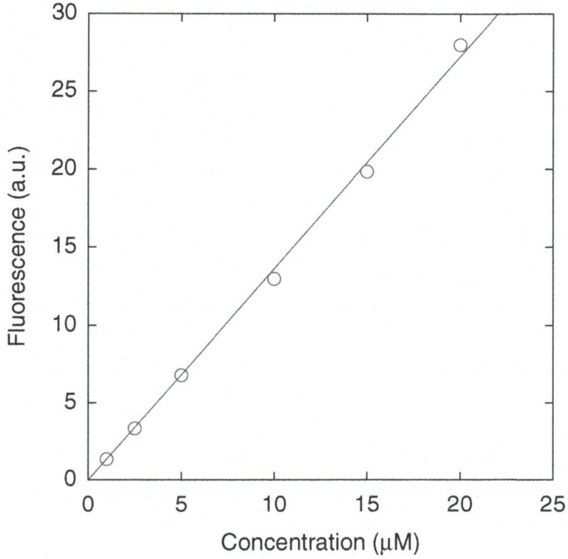

Fig. 7.13 Calibration curve used for SOD quantification. Experimental conditions: $\lambda_{exc} = 260$ nm; emission cutoff filter = 295 nm; Bandwidth 5 nm in both excitation and emission; cell size = 5 mm exc. side, 10 mm emission side

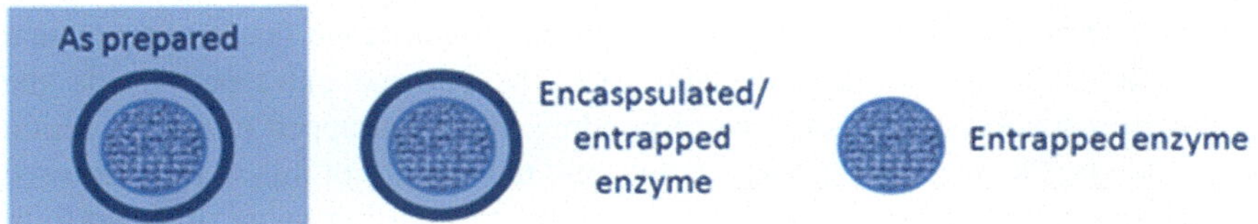

Fig. 7.14 Schematic representation of the different steps of the experiment. The *dark blue circle* represents the lipid bilayer, while the *light blue grid* represents the polymeric nanoparticle

7.8 Quantification of Entrapped Enzyme

To quantitatively characterize the fraction of enzyme entrapped within the NPs, SOD intrinsic fluorescence was exploited. The quantification was obtained by the acquisition of a calibration curve (Fig. 7.13), collecting spectra of different amounts of SOD in buffer. The different steps in which SOD has been quantified are reported in Fig. 7.14. At each step both the enzyme concentration (as provided by the fluorescence intensity) and its activity were determined.

For these experiments, two stocks of NPs with a nominal diameter of 400 nm were prepared; in the first, a 4 mg/ml solution of SOD was included, while the second was synthesized without the enzyme; the spectra accquired from the second solution were used for the background subtraction in the fluorescence spectra.

Before any purification, a spectrum was acquired from the stock solution. NPs in this stage are referred to as *"as prepared"*. In this case, the value of SOD concentration is known from the conditions used in the synthesis (12.8 μM), but it was anyway assessed with fluorescence, resulting to be 12.6 μM. The enzyme activity of this sample was 630 units/ml. This value is due only to the free, unentrapped enzyme that is in the water volume outside the vesicles, because the substrate diffusion inside the liposomes is strongly inhibited by the membrane. When the bilayers were broken by SDS addition, the activity rose to 980 units/ml, due to the enzyme molecules previously enclosed in the membrane, or entrapped in the hydrogel network. The presence of SDS does not interfere with the activity of the enzyme [17].

Subsequently, a first dialysis was carried out, maintaining, in this case, the membrane integrity. At this point, a further fluorescence spectrum was collected; in this case, the signal is given by the *"encapsulated"* enzyme, which is inside the lipid vesicles. The spectrum obtained from this experiment indicated that the encapsulated enzyme fraction corresponded to 37 % of the total SOD amount. The activity of this sample was practically not detectable, confirming that free SOD was efficiently removed from the NPs suspension. After the addition of SDS, the value of activity resulted to be 170 units/ml, corresponding to about 17 % of the starting value.

After the quantification of the encapsulated enzyme fraction the sample was submitted to a further dialysis cycle, to allow the removal of all the enzyme that was initially located inside the lipid vesicle but was not entrapped in the hydrogel network. Thus, a further spectrum was acquired, in which the fluorescence signal was given only by *"entrapped"* SOD, i.e. the enzyme entrapped inside the polymeric network. The amount corresponding to the entrapped enzyme fraction resulted to be the 30 % of the total amount and the activity of this sample featured a value of 45.5 μ/ml, corresponding to 4.5 % of the starting value. It is worth to underline that this purification allowed the removal of the protein mobile fraction that was evidenced by the FRAP measurements.

The sample was then purified again, with multiple dialysis treatments, to check for a very slow release of the entrapped enzyme. The fluorescence and the enzymatic activity were measured again every 24 h, revealing a progressive decrease. This means that a slow release of the enzyme from the NPs occurs after the membrane removal step. The fluorescence spectra collected for every step, and the corresponding percentages are reported in Figs. 7.15 and 7.16. The activity values are summarized in Fig. 7.17.

The fluorescence experiments illustrated that after the lipid membrane removal, the fraction of entrapped enzyme slightly decreases after each purification cycle. This indicates that when the membrane is solubilized, a slow enzyme release starts, and that several days are needed for a complete release.

Another important observation is that the fluorescence and activity values measured for the same sample in the same steps of this experiment resulted to be significantly different. A comparison is reported in Table 7.3.

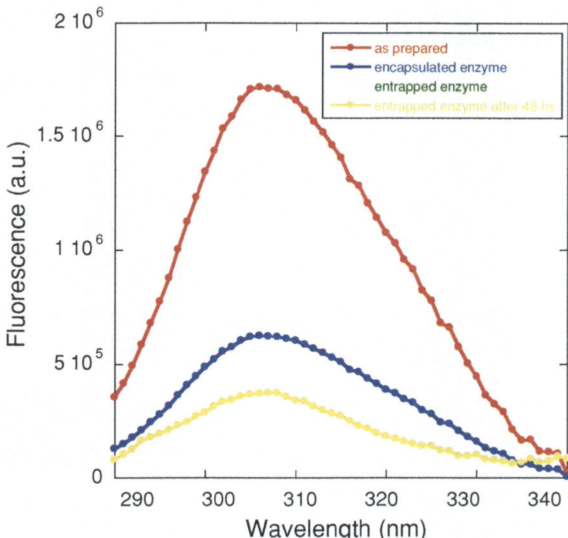

Fig. 7.15 SOD emission spectra of the unpurified NPs solution (*red*), of the encapsulated enzyme (*blue*), and of the entrapped enzyme after 24 h (*yellow*). Experimental conditions: $\lambda_{exc} = 260$ nm; emission cutoff filter = 295 nm; bandwidth: 5 nm both in excitation and emission; cell size = 5 mm in the excitation side, 10 mm in the emission side

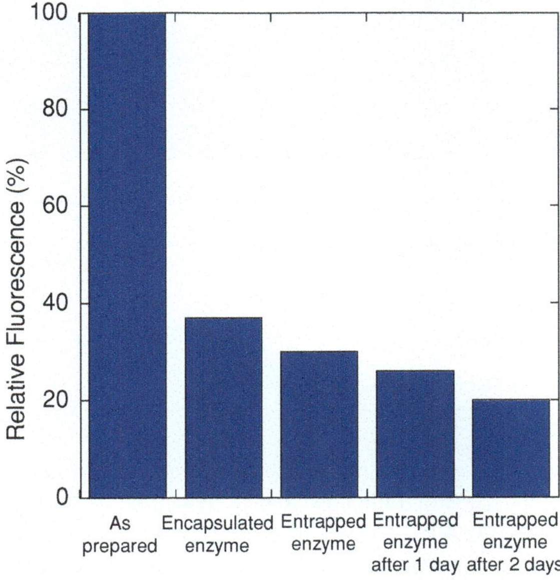

Fig. 7.16 Fluorescence of SOD inside NPs as a percentage of the bulk solution signal

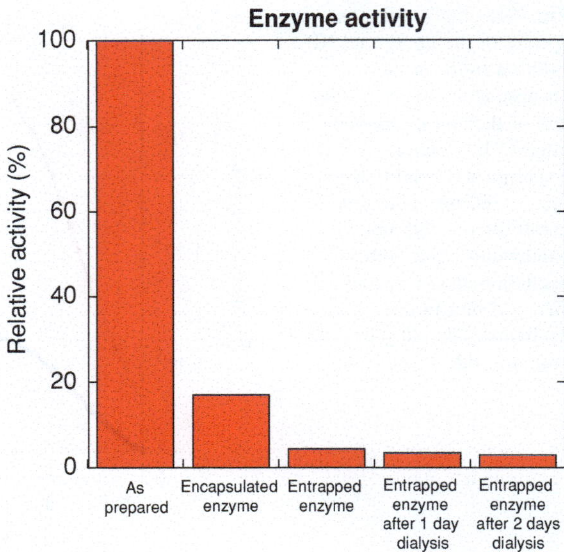

Fig. 7.17 Enzymatic activity of SOD inside NPs with respect to the activity of the unpurified solution

Table 7.3 Summary of the amount and activity of SOD inside NPs as a function of days. F(%) is the fluorescence, A (%) is the activity

Sample	F (%)	A (%)	Concentration (μM) from fluorescence spectrum	Specific activity (u/mg)
Encapsulated enzyme	37	17	4.7	2300
Entrapped enzyme	30	4.4	3.8	800
Entrapped enzyme after 24 hours	29	3.4	3.7	600
Entrapped enzyme after 48 hours	21	3.3	2.7	700

This result indicates that a reduction of the specific activity of SOD occurs inside the NPs. The specific activity of an enzyme is defined as the number of units per mg. In this case, this parameter has been calculated using the concentration obtained from the fluorescence experiment. The specific activity of the enzyme in the free solution resulted to be $A_0 = 8620$ units/mg. After the inclusion inside the NPs, the specific activity of SOD showed a significant reduction, down to 2297 units/mg, i.e. 27 % of the initial value.

A first explanation of these results could involve a partial denaturation of the enzyme inside the NPs. Anyway, the mild synthesis conditions make this

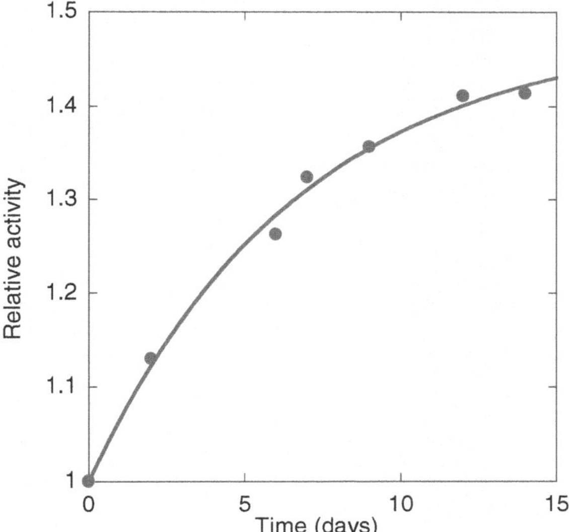

Fig. 7.18 Increase of activity inside NPs as a function of time followed during the days after preparation

supposition quite unlikely. Another hypothesis could involve a different enzyme conformation inside the polymeric network [19], or simply a slower diffusion of the substrates within the hydrogel. To clarify this point, the activity of the same sample was measured right after the preparation, and then, during the following days, to check whether a variation of activity occurs as a consequence of the partial enzyme release from the NPs (shown by the fluorescence experiment), i.e. if the enzyme released in the free solution is more active than the entrapped one. The sample was not submitted to any purification process, and the released protein remained within the solution.

In this case, there is an evident activity increase with time (Fig. 7.18). This means that the fraction of released enzyme recovers its activity in the free solution, at least partially. Indeed, the activity value recorded during the days is given by the released fraction and by the fraction that is still entrapped inside the gel network. The calculated time constant for the process resulted to be $\tau = 6.8$ days.

The reduced activity of SOD entrapped inside the NPs does not preclude their therapeutic application because the enzyme is still active. The enzyme release form NPs could be obviously a problem for the shelf-life of the drug, but the presence of the membrane can be used to keep the enzyme within the gel during storage, and it can be removed immediately before the use of NPs.

It is worth to underline that the quantity of entrapped enzyme can be obviously increased by synthesizing larger liposomes, but the size of NPs also plays a key role in their biodistribution and permanence inside the organisms. It has been demonstrated that the longest blood residence time is given by NPs characterized by a diameter of about 100 nm, while NPs with a size around 400–500 nm accumulate within liver and spleen. The best dimension for a long-circulating

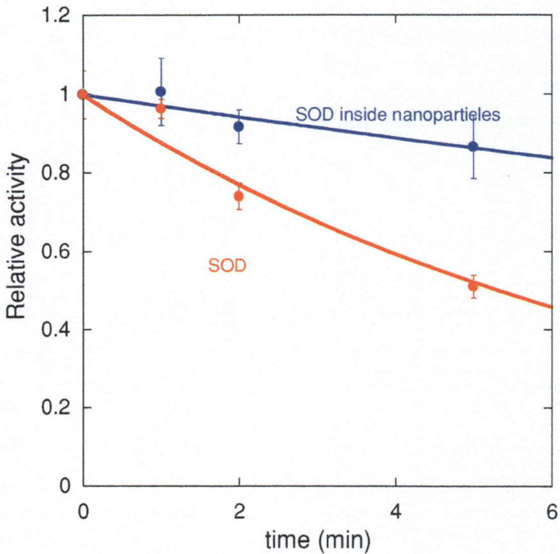

Fig. 7.19 SOD thermal inactivation. Enzymatic activity of free SOD (*blue*) and of entrapped SOD (*red line*). The heating temperature was 80 °C, and the activity was measured on each sample after cooling in an ice bath

nanoparticle is less than 200 nm [1]. In this work the size of NPs is larger to allow the entrapment of a consistent enzyme fraction, as a proof of principle of the feasibility of the system. However, the vehicle dimensions must be chosen according to the planned therapeutic purposes.

7.9 Enzyme Stabilization

Protein stability is a fundamental requirement of a therapeutic formulation.

In the specific case of SOD, the enzyme is inactivated at high temperature, because of the formation of aggregated structures, which are insoluble and precipitate [11, 12, 15].

The polyacrylamide gel represents a confining environment for entrapped molecules. The confinement within the gel should provide a stabilization of proteins, by decreasing the entropy of the unfolded state through an excluded volume effect [6]. In particular, it has been seen that the thermal stability of proteins is enhanced in a microporous polyacrylamide gel (with a pore size of 10 nm or less) environment [6]. Thermal inactivation experiments were carried out on free SOD and on SOD-containing NPs. Samples were heated at 80 °C and the enzyme activity was tested at different times. As reported in Fig. 7.19, in the case of free SOD more than 50 % of activity was lost in a few minutes. The activity decrease was much less evident in the case of entrapped SOD: in this case, the reduction was only about the 6 % of initial activity.

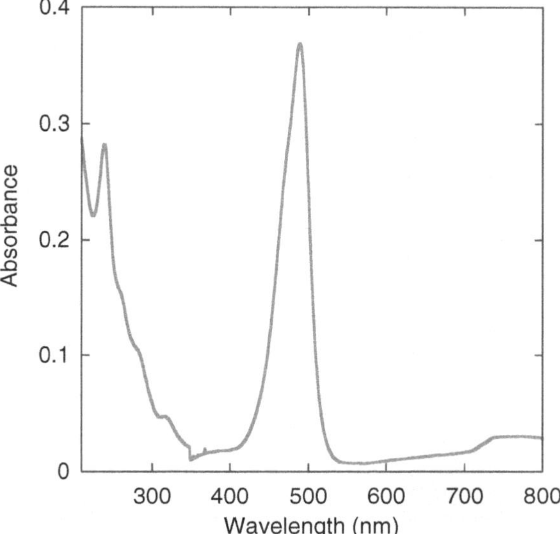

Fig. 7.20 UV-visible spectrum of a 125 μM solution of FLAC. Pathlength: 1 cm

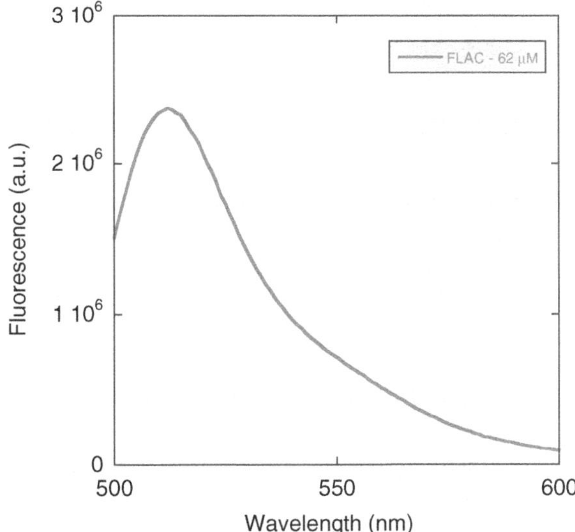

Fig. 7.21 *Fluorescence* spectrum of a 0.62 μM FLAC solution. Experimental conditions: $\lambda_{exc} = 490$ nm; slits 2/2; cell size = 5 mm excitation side, 10 mm emission side

7.10 FLAC Synthesis and Characterization

In order to visualize NPs inside the cells to verify their uptake, FLAC has been synthesized, as described in Sect. 6.10, and used to label the NPs.

UV-visible and fluorescence spectra of FLAC were collected, showing the characteristic absorption and emission bands of the fluorescein chromophore. The absorption spectrum is reported in Fig. 7.20, while the fluorescence spectrum is reported in Fig. 7.21.

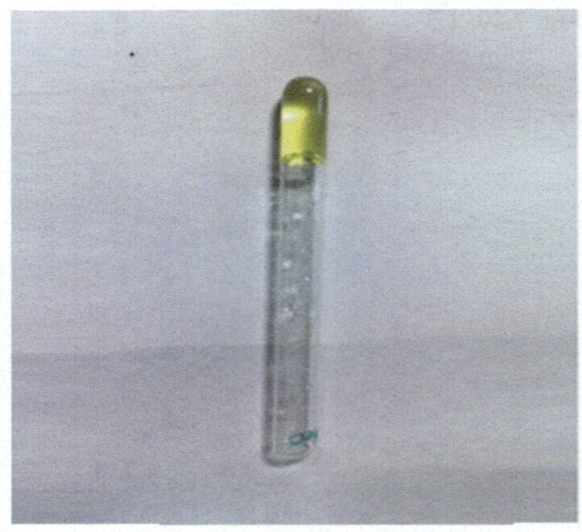

Fig. 7.22 Bulk polymerization of fluorescent gel (T = 20 %, C = 5 %)

Table 7.4 Composition of the hydrogel used for fluorescent NPs synthesis

Reagent	mg/ml	Molarity
Acrylamide	130.1	2.674
MBA	10.4	0.0674
FLAC	0.05	1.124e-04

7.11 Gel Polymerization in the Presence of FLAC

The presence of FLAC during polymerization causes a strong absorption of the UV light used during sample irradiation, possibly inhibiting the photopolymerization process. To rule out this possibility, the gelation process was checked in bulk hydrogel samples.

A first gel was prepared with the T = 20 %, C = 5 % formulation, using a FLAC to acrylamide molar ratio of 1. In this case, the polymerization did not occur. The FLAC concentration was then decreased 10-fold. In this case, the polymerization was incomplete. The maximum quantity of FLAC to allow a complete polymerization resulted to be the 0.0047 % with respect to the monomer (Fig. 7.22). NPs were synthesized using the formulation described in Table 7.4. The fluorescence signal from the NPs sample resulted to be good enough to allow the cell-internalization investigation.

7.12 Cell Uptake Experiments

For cell uptake experiments, murine fibroblasts were incubated with a 0.4 nM solution of fluorescent NPs. Both membrane-coated and naked NPs were efficiently internalized by cells already after 24 h (Figs. 7.23 and 7.24), as shown by the green fluorescence signal within cells.

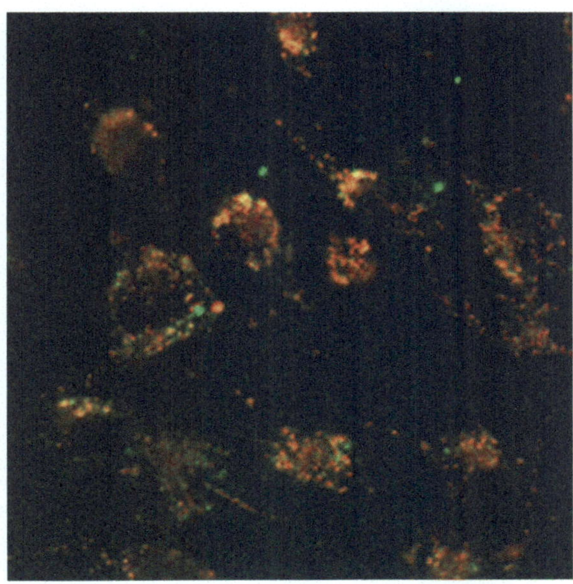

Fig. 7.23 Confocal image of a fibroblast. after a 72 h incubation with a solution of membrane-coated NPs

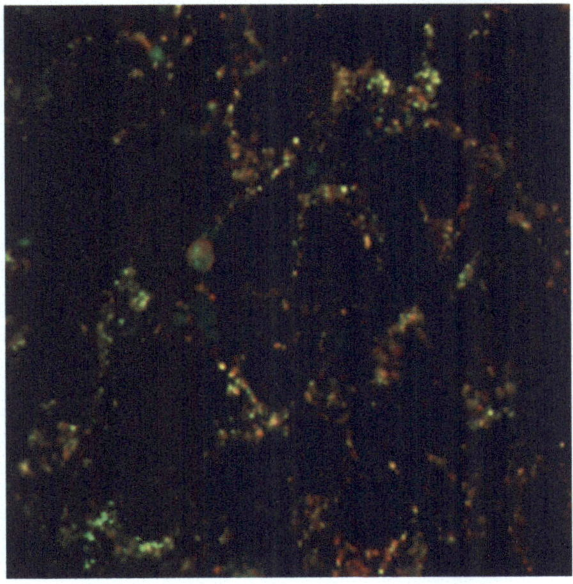

Fig. 7.24 Confocal image of fibroblasts after a 24 h incubation with a solution of naked NPs

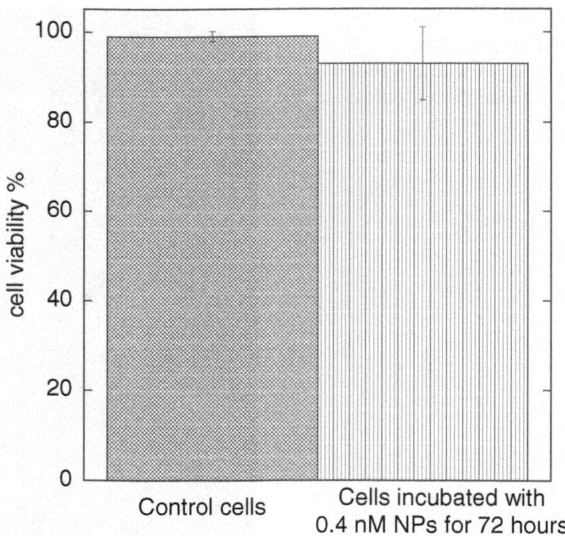

Fig. 7.25 Trypan blue assay results of cell viability after 72 h of incubation with a 0.4 nM NPs solution

7.13 Cell Viability Assays

The biocompatibility of NPs is a fundamental requirement for their therapeutic application. Acrylamide-based gels biocompatibility has been demonstrated by several studies [14, 20, 23]; nevertheless, the potential toxicity of the samples used in this work was tested on cell cultures.

7.13.1 Trypan Blue Assay

The trypan blue assay is a test of the cell membrane integrity; this is a fundamental requirement for a living cell, and thus it can be used indirectly as a test for viability. In this case, the results obtained in the presence of NPs and for the control culture were very similar (Fig. 7.25). This result means that NPs do not interfere with the cell integrity.

7.13.2 Live or Dead Assay

The live or dead assay provided further information about the cell membrane integrity. In this case, the experiments showed again that, after a 48 h incubation, no relevant differences occur between the control cell culture (Fig. 7.26) and the cell culture treated with NPs (Fig. 7.27).

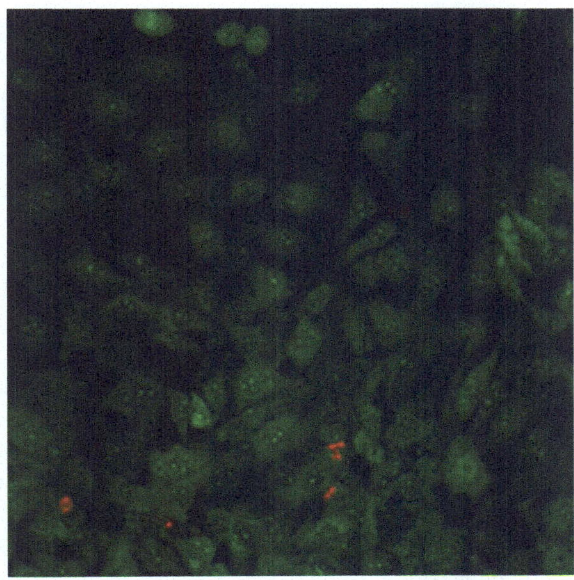

Fig. 7.26 Confocal image of the control cell culture in the live or dead assay

Fig. 7.27 Confocal image of the NPs-treated cell culture in the live or dead assay after 48 h of incubation

It is worth to underline that all the cell viability assays have been performed using NPs which did not contain SOD, to detect exclusively the effects of the vehicle on cell cultures.

7.14 Discussion

The huge therapeutic potential of enzymes is still poorly exploited, due to the many problems involved in their administration and delivery. In this work, a novel approach has been presented that can contribute to solve some problems related to the short half-life of these molecules, as well as to the raise of immunogenic response.

By contrast to previously proposed enzyme delivery vehicles, which are designed to release the enzyme, so that it can exert its activity outside the carrier, the NPs presented in this work have been designed to obtain a system that retains the enzyme, allowing it to maintain its catalytic function without being released. This protects the protein against antibodies, degradation and denaturing conditions, enhancing the half-life of the molecule during the entire period of the therapy.

Another important difference between the NPs presented in this work and all the most used micro- and nano-particle formulations is that the latter involve denaturing conditions during their synthesis. By contrast, the use of liposomes as nanotemplates allows the use of physiologic conditions in the entire process, and an easy selection of the appropriate dimensions for NPs through the extrusion step.

The synthesized NPs have a reasonable entrapping efficiency, and a good cell uptake. In addition, their biocompatibility, evaluated with different assays, resulted to be very good.

All these characteristics make the NPs presented in this part of the present thesis a promising system for enzyme-based therapies.

References

1. Alexis F, Pridgen E, Molnar LK, Farokhzad OC (2008) Factors affecting the clearance and biodistribution of polymeric nanoparticles. Mol Pharm 5:505–515
2. Asnaghi D, Giglio M, Bossi A, Righetti PG (1997) Quasi-ordered structures in highly cross-linked poly(acrylamide) gels. Macromolecules 30:6194–6198
3. Axelrod D, Koppel DE, Schlessinger J, Elson J, Webb WW (1976) Mobility measurement by analysis of fluorescence photobleaching recovery kinetics. Biophys J 16:1055–1069
4. Baselga J, Hernansez-Fuentes I, Pierola IF, Llorente MA (1987) Elastic properties of highly-crosslinked polyacrylamide gels. Macromolecules 20:3060–3065
5. Baumann G, Chrambach A (1976) A highly crosslinked, transparent polyacrylamide gel with improved thermal stability for use in isoelectric focusing and isotachophoresis. Anal Biochem 70:32–38
6. Bolis D, Politou AS, Kelly G, Pastore A, Temussi PA (2004) Protein stability in nanocages: a novel approach for influencing protein stability by molecular confinement. J Mol Biol 336:203–212
7. Boyde TRC (1976) Swelling and contraction of polyacrylamide gel slabs in aqueous solution. J Chromatogr 124:219–230

8. Braeckmans K, Peeters L, Sanders NN, De Smedt SC, Demeester J (2003) Three-dimensional fluorescence recovery after photobleaching with the confocal scanning laser microscope. Biophys J 85:2240–2252
9. Brannon-Peppas L, Peppas NA (1990) The equilibrium swelling behavior of porous and non-porous hydrogels. Absorbent polymer technology, Stud Polym Sci 8:67–102
10. Eichenbaum GM, Kiser PF, Dobrynin AV, Simon AS, Needham D (1999) Investigation of the swelling response and loading of ionic microgels with drugs and proteins: the dependence on cross-link density. Macromolecules 32:4867–4878
11. Folcarelli S, Battistoni A, Carrì MT, Polticelli F, falconi M, Nicolini L, Stella L, Rosato N, Rotilio G, Desideri A (1996) Effect of Lys-Arg mutation on the thermal stability of superoxide dismutase: influence on the monomer-dimer equilibrium. Prot Eng 9:323–325
12. Forman HJ, Fridovich I (1973) On the stability of superoxide dismutase. J Biol Chem 248:2645–2649
13. Ganji F, Vashegani-Farahani S, Vashegani-Farahani E (2010) Theoretical description of hydrogel swelling: a review. Iran Polym J 19:375–398
14. Karadag E, Saraydin D, Cetinkaya S, Guven O (1996) In vitro swelling studies and preliminary biocompatibility evaluation of acrylamide-based hydrogels. Biomaterials 17:67–70
15. Lepock JR, Frey HE, Hallewell RA (1990) Contribution of conformational stability and reversibility of unfolding to the increased thermostability of human and bovine superoxide dismutase mutated at free cysteines. J Biol Chem 265:21612–21618
16. Lira LM, Martins KA, Cordoba de Torresi SI (2009) Structural parameters of polyacrylamide hydrogels obtained by the equilibrium swelling theory. Eur Polym J 45:1232–1238
17. Marklund SL (1984) Properties of extracellular superoxide dismutase from human lung. Biochem J 220:269–272
18. Nagash HJ, Okay O (1996) Formation and structure of polyacrylamide gels. J Appl Polym Sci 971–979
19. Pastor I, Pietro M, Mateo CR (2008) Effect of sol-gel confinement on the structural dynamics of the enzyme bovine Cu, Zn superoxide dismutase. J Phys Chem 112:15021–15028
20. Peppas NA, Huang Y, Torres-Lugo M, Ward JH, Zhang J (2000) Physicochemical foundations and structural design of hydrogels in medicine and biology. Annu Rev Biomed Eng 2:9–29
21. Russel SM, Carta G (2005) Mesh size of charged polyacrylamide hydrogels from partitioning measurements. Ind Eng Chem Res 44:8213–8217
22. Salin ML, Wilson WW (1981) Porcine superoxide dismutase. Isolation and characterization of a relatively basic cuprozinc enzyme. Mol Cell Biochem 36:157–161
23. Saraydin D, Ünver-Saraydin S, Karadag E, Koptagel E, Güven O (2004) In vivo biocompatibility of radiation crosslinked acrylamide copolymers. Nucl Instr Meth B 217:281–292
24. Song KB, Damodaran S (1991) Influence of electrostatic forces on the adsorption of succinylated beta-lactoglobulin at the air-water interface. Langmuir 7:2737–2742
25. Soumpasis DM (1983) Theoretical analysis of fluorescence photobleaching recovery experiments. Biophys J 41:95–97

Concluding Remarks

In this thesis two main topics have been addressed, i.e. the interaction of AMPs with lipid membranes, and an innovative approach, employing phospholipid vesicles for the synthesis of NPs entrapping the enzyme SOD, aimed at therapeutic applications. These subjects are in different ways related to serious threats to human health that became alarming in the last years, i.e. the bacterial resistance to traditional antibiotics, and pathologies deriving from oxidative stress. In the investigation of both these issues phospholipid membranes had a central role.

In the first part of this work model membranes, constituted mainly by liposomes, or by lipid bilayers deposited on a support, have been exploited to mimic biological membranes. These model systems allow the physico-chemical investigation of the interaction between peptides and membranes, avoiding interference from other components present in real biological membranes, such as membrane proteins or other nonlipidic components. In this thesis, model membranes have been used to elucidate the membrane-perturbation properties of AMPs.

The mechanism of pore formation of the peptide GAIV has been completely elucidated, clarifying a mode of action that had been previously heatedly debated in the literature, and contributing to solve an apparent contradiction due to the small length of this peptide, which is not sufficient to span the entire bilayer but is apparently able to form transmembrane pores. The results obtained for GAIV showed that an AMP can cause a strong deformation of the lipid bilayer, thus obtaining a match between its length and the thickness of the membrane, and allowing the formation of transmembrane pores. The physico-chemical driving forces of this process were clarified as well. This important goal has been reached with the use of a combination of techniques, i.e. fluorescence and neutron spectroscopy, and molecular dynamics simulations, that resulted to be particularly indicated to investigate the mechanism of interaction between membranes and AMPs. This approach can be successfully applied to the study of analogous systems. These results are important also because the mechanism assessed for GAIV can be relevant also for several other short peptaibols, which have very similar structures.

S. Bobone, *Peptide and Protein Interaction with Membrane Systems*,
Springer Theses, DOI: 10.1007/978-3-319-06434-5,

Another crucial point for the rational design of AMPs and antimicrobial peptidomimetics is a detailed understanding of the structural properties leading to cell selectivity. The ability to distinguish between the bacterial and the host cells is one of the most interesting features of AMPs, which makes them promising candidates as a new class of antibiotics. In this thesis, the role of the central Pro residue, which is common to many AMPs, has been investigated, revealing that it is a fundamental requirement for selectivity. The data presented here indicate that Pro its removal causes the stabilization of the helical structure of the peptide in water, increasing the peptide hydrophobicity by exposing its apolar residues, and, as a consequence, its affinity towards neutral membranes that mimic those of eukaryotic cells. These results have been obtained for a model peptide, P5, but lead to a more general statement, i.e. that the role of central Pro is fundamental in the inhibition of peptide binding to neutral membranes, like those of mammalian cells, and that this feature must be taken into account in the development of new peptide-based bactericidal molecules.

The last investigation presented on AMPs involves the possibility to modulate the biological activity of a peptide. It has been shown that a CPP can become an AMP by operating just slight modifications in its amino acidic sequence. This is due to the fact that, even though biological activities of CPPs and AMPs are quite different, their structures have several common features. The dramatic change in the activity caused by minimal modifications in the sequence of a peptide illustrates the difficulties involved in the rational design of membrane-active peptides. On the other hand, the data presented in this thesis demonstrated that the feasibility of a peptide endowed with both antibacterial and penetrating activity and this could be useful for the development of new antibiotic drugs, and for several other therapeutic applications. For instance, it has been shown in the last years that some cationic AMPs are also specifically active against tumor cells; thus, the delivery of a specific drug can be enhanced with the use of such a peptide as a carrier system.

All the presented studies on AMPs definitely give a contribution to a deeper understanding of their behavior and structural characteristics, and therefore to the development of novel peptide-based drugs.

In the second part of this work another important characteristic of lipid vesicles has been exploited, i.e. the capability to create in their inner volume a segregated environment, in which a chemical reaction can occur. This environment has a nanometric size and in this case it has been used to obtain a nanoparticle by a photopolymerization reaction that has been carried out only inside the vesicles, and avoided in the bulk solution. In this way, it has been possible to obtain hydrogel NPs in physiological, mild conditions; this characteristic makes them suitable to entrap enzymes preserving their catalytic activity. The available formulations of enzyme vehicles are designed to release the drugs to the tissues; the goal of this work has instead been the development of a system in which an enzyme can exert its function without being released and, therefore, without being exposed to proteolytic enzymes or to antibodies. The enzyme chosen for this study is SOD. The gel synthesis protocol has been optimized to ensure the entrapment of enzyme molecules, while allowing the diffusion of smaller molecules, like substrates and products, so that the enzyme can maintain its biological function inside the NPs.

The anti-oxidant function of SOD has been poorly exploited in therapeutic applications, despite its relevance as a radical scavenger, due to problems related mainly to immunogenic response, that can be solved using the approach presented in this thesis. The increasing exposure to ROS deriving from ionizing radiations and air pollutants makes these results even more important, opening a new way to the development of a SOD-based drug, and more generally of therapies based on any enzyme with substrates and products small enough to diffuse across the gel network.

Presentations to Scientific Congresses Derived from this Thesis

- SIBPA 2012 Congress, (Ferrara, Italy), September 17–20. Poster presentation. *Membrane thickness and the mechanism of action of the short peptaibol trichogin GA IV.*
- 32 EPS Meeting, Athens (Greece), September 1–7, 2012 Oral communication in the Dr. Scham's Young Investigator Contest. *Membrane thickness and the mechanism of action of the short peptaibol trichogin GA IV.* Proceedings in Peptides 2012 (pp. 138–139). Poster presentation: Membrane-perturbing effects of antimicrobial peptides: a systematic spectroscopic analysis. Proceedings in Peptides 2012 (pp. 166–167).
- 13th Naples Workshop on Bioactive peptides. Naples (Italy), June 2/7, 2012. Poster presentation: *Membrane thickness and the mechanism of action of the short peptaibol Trichogin GA IV.*
- New Antimicrobials. Trieste (Italy), May 25–26, 2012. Poster presentation: *Membrane thickness and the mechanism of action of the short peptaibol Trichogin GA IV.*
- Nanodrug Delivery from the Bench to the Patient. Rome, ISS (Italy), October 10–13, 2011. Poster Presentation: *Hydrogel nanoparticles for enzyme-based therapies.*
- Biophysical Meeting 2011, Baltimore (USA), March 4–10, 2011. Poster presentation: *The importance of being kinked: role of pro residues in the selectivity of helical antimicrobial peptides.*
- SIBPA 2010 Congress, Arcidosso (Italy), September 11–14, 2010. Oral communication: *Hydrogel nanoparticles for enzyme-based therapies.* Poster presentation: *The importance of being kinked: role of pro residues in the selectivity of helical antimicrobial peptides.*
- 31st European Peptide Symposium, Copenhagen (Denmark), September 5–9, 2010. Poster Presentation: *Effect of helix kink on the activity and selectivity of an antimicrobial peptide* and *Membrane insertion of para-cyanophenylalanine-labeled alamethicin analogues. Correlation of fluorescence and infrared absorption data.* Proceedings in Peptides 2010, pp. 370–373.

S. Bobone, *Peptide and Protein Interaction with Membrane Systems*,
Springer Theses, DOI: 10.1007/978-3-319-06434-5,

- 12th Naples Workshop on Bioactive Peptides and 2nd Italy-Korea Symposium on Antimicrobial Peptides, Naples (Italy), June 4–7, 2010. Oral communication in the Murray Goodman Young Investigators' Session. *The importance of being kinked: role of pro residues in the selectivity of helical antimicrobial peptides.* Poster presentation: *Membrane insertion of para-cyanophenylalanine-labeled alamethicin analogues. Correlation of fluorescence and infrared absorption data.*
- European Biophysics Congress, Genoa (Italy), July 11–15, 2009. *Effect of helix kink on the activity and selectivity of an antimicrobial peptide.*
- 1st Italy-Korea Symposium, Gwangju (Korea), July 24–25, 2008. *Membrane perturbing activity of a new analogue of the cell-penetrating peptide Pep-1* and *Effect of helix kink in the activity and selectivity of an antimicrobial peptide.* 11th Naples Workshop on Bioactive Peptides, Napoli, Italy, May 24–27, 2008.

The manufacturer's authorised representative in the EU is Springer Nature Customer Service Centre GmbH, Europaplatz 3, 69115 Heidelberg, Germany. If you have any concerns regarding our products, please contact ProductSafety@springernature.com

Printed and bound by CPI Group (UK) Ltd, Croydon, CR0 4YY
15/07/2026
02167645-0008